2012 NATIONAL GREEN BUILDING STANDARD™

ICC 700-2012

Approved American National Standard
ANSI

NAHB
National Association of Home Builders

ICC
INTERNATIONAL CODE COUNCIL

2012

ICC 700
National Green Building Standard™

First Printing: January 2013

ISBN-13: 978-0-86718-697-0
eISBN-13: 978-0-86718-716-8

Disclaimer
This publication provides accurate information on the subject matter covered. The publisher is selling it with the understanding that the publisher is not providing legal, accounting, or other professional service. If you need legal advice or other expert assistance, obtain the services of a qualified professional experienced in the subject matter involved. The NAHB has used commercially reasonable efforts to ensure that the contents of this volume are complete and appear without error; however the NAHB makes no representations or warranties regarding the accuracy and completeness of this document's contents. The NAHB specifically disclaims any implied warranties of merchantability or fitness for a particular purpose. The NAHB shall not be liable for any loss of profit or any other commercial damages, including but not limited to incidental, special, consequential or other damages. Reference herein to any specific commercial products, process, or service by trade name, trademark, manufacturer, or otherwise does not necessarily constitute or imply its endorsement, recommendation, or favored status by the NAHB. The views and opinions of the author expressed in this publication do not necessarily state or reflect those of the NAHB, and they shall not be used to advertise or endorse a product.

Trademarks: "National Association of Home Builders," "NAHB," "BuilderBooks," and the "NAHB" logo are registered trademarks of the National Association of Home Builders of the United States. "National Green Building Standard" is a trademark of the National Association of Home Builders of the United States.

Trademarks: "ICC International Code Council," and the "International Code Council" logo are registered trademarks of the International Code Council, Inc.

PRINTED IN THE U.S.A.

PREFACE

Introduction

Green buildings are designed, constructed, and operated with a goal of minimizing their environmental footprint. In both new construction and renovation, the building and its site are designed in an integrated manner using environmentally preferable practices and materials from start to finish. Many green features also carry direct consumer benefits such as lower monthly utility bills, greater comfort, reduced maintenance, and increased value.

The ICC 700 *National Green Building Standard*™ (NGBS) is the premier rating system for sustainable construction that sets no limit on sustainable practices – while reflecting the realities of budgets, geographical constraints, and local market requirements.

In 2007 the National Association of Home Builders (NAHB) and the International Code Council (ICC) partnered to establish this much-needed and nationally recognizable green building and rating system that culminated in the publication of the 2008 NGBS.

This edition is a revised, expanded version of that standard, completed in 2012 to acknowledge advances in building science, reflect recent model code improvements and add more choices for compliance.

The 2012 NGBS also incorporates improvements and enhancements gleaned from the standard's first four years of implementation. But one important feature has not changed: NGBS remains the first green building rating system to undergo the full consensus process and receive approval from the American National Standards Institute (ANSI) and the only residential system to do so. NGBS provides practices for the design, construction, occupation and maintenance of green buildings, which can be used as a measurement of environmental performance levels in nationally recognized certification programs for single-family homes, low-, mid-, and high-rise multifamily buildings, residential remodeling, and site development projects.

Using a points-based system, a home or building can attain one of four rating levels — Bronze, Silver, Gold, or Emerald — depending on the green practices included. For a building to attain any certification level, all applicable mandatory provisions must be correctly implemented. NGBS also requires that the builder or remodeler incorporate a minimum number of features in each of six categories for each rating level (lot development, resource efficiency, energy efficiency, water efficiency, indoor environmental quality, and home owner education).

As an ANSI-approved green building rating system, the NGBS provides builders with a credible definition of green building and development and a useful measurement of its relative environmental performance. Because it provides a flexible, expansive point-based system for certification, it also offers builders a process to build affordable green homes, as appropriate for varying climates, and meets the needs of the market and home buyers.

2012 NGBS Development

The Consensus Committee on the *National Green Building Standard*™, consisting of 41 members, was assembled of those entities and interests that are affected by the provisions of the Standard. In addition, a total of seven Task Groups were formed by the specific area of technical expertise to serve as a resource to the Consensus Committee. The Task Groups included committee members as well as other industry experts. The entire Standard was open for public to submit proposed changes before the Consensus Committee and the Task Groups began its work on revising and expanding the provisions of the Standard. The Consensus Committee met three times during 2011 and 2012 to discuss and take formal actions first on proposed changes and then on public comments. All meeting were open to the public to provide an opportunity to testify before the Consensus Committee. All committee actions were also balloted through formal letter ballots. Overall, the Consensus Committee reviewed and acted upon over 800 proposed changes and public comments ranging from revisions to individual provisions to addition of entire chapters. The scope of the Standard was expanded to include accessory structures.

ANSI Approval

The ICC 700-2012 *National Green Building Standard*™ was approved by the American National Standards Institute (ANSI) as an American National Standard on January 10, 2013.

Maintenance

This Standard was developed by the NAHB Research Center, an ANSI Accredited Standards Developer, and is revised as required by ANSI. Proposals for revising this edition of the *National Green Building Standard*™ are welcome. Please visit the NAHB Research Center website (www.nahbrc.com/NGBS) for a proposal form and instructions.

Disclaimer

NAHB Research Center, NAHB, ICC, their members and those participating in the development of this Standard accept no liability resulting from compliance or noncompliance with the provisions. Neither NAHB nor ICC nor the NAHB Research Center has the power or authority to enforce compliance with the contents of this Standard. Similarly, neither NAHB nor ICC make any representations or warranties regarding enforcement, application or compliance with this Standard or any part thereof.

2012 Consensus Committee on the National Green Building Standard™

At the time of ANSI approval, the Consensus Committee consisted of the following members:

Chair	Donald L. Pratt
Vice Chair	Ray Tonjes
Committee Staff	Vladimir Kochkin
ICC Staff Liaison	Allan Bilka

Committee Member	Representative
Air Barrier Association of America, Inc. (P)	Laverne Dalgleish
Air-Conditioning, Heating and Refrigeration Institute (P)	Frank A. Stanonik
American Gas Association (P)	Ted Arthur Williams
	Paul Cabot (Alt.)
American Institute of Architects (U)	David S. Collins, FAIA
American Wood Council (P)	Kenneth Bland
	Sam W. Francis (Alt.)
Association of Home Appliance Manufacturers (P)	Matthew Brian Williams
BME Associates (U)	Bruce G. Boncke
BOMA International (U)	Ron Burton
C. F. Evans & Company (U)	Patrick Westbury
	Joel Freeman (Alt.)
CECS of MI, LLC (U)	Donald L. Pratt
City of Denton, TX (G)	Kurt Spence Hansen
City of Keene, NH (G)	Medard Kopczynski
City of Longmont, CO (G)	Chris Allison
City of Scottsdale, AZ (G)	Anthony C. Floyd
ConSol (U)	Michael G. Hodgson
Edison Electric Institute (P)	Steven Rosenstock
Environmental Solutions Group (U)	Steven Armstrong
Forest City Land Group (U)	William Sanderson
GREENGUARD Environmental Institute (P)	Josh Jacobs
Habitat for Humanity International (U)	Matt Clark
	Mike Mongeon (Alt.)
Memphis Land Bank, Inc. (G)	Molly A. Beard
National Multi Housing Council (U)	Paula Cino
	Ron Nickson (Alt.)
North American Insulation Manufacturers Association (P)	Charles C. Cottrell
	Darrell K. Winters (Alt.)
NVR Inc. (U)	Dan Simon
Plastic Pipe and Fittings Association (P)	Michael William Cudahy
Portland Cement Association (P)	Donn C. Thompson, AIA, CGP, LEED
	Stephen V. Skalko, P.E. (Alt.)
Ray Tonjes Builder, Inc. (U)	Ray Tonjes

State of California – Dept. of Housing and Community Development (G) .... Doug Hensel

State of New Jersey – Division of Codes and Standards (G) .... Darren Molnar-Port

Steel Framing Alliance (P) .... Maribeth S. Rizzuto
Mark Nowak (Alt.)

Steve Easley & Associates (U) .... Steve Easley

The Sullivan Company, Inc. (U) .... Paul L. Sullivan, CGR, CAPS

U.S. Army (G) .... Deborah Reynolds

U.S. Department of Energy (G) .... Jeremiah L. Williams

U.S. Department of Housing & Urban Development (G) .... Dana Bres, P.E.
Mike Blanford (Alt.)

U.S. Environmental Protection Agency (G) .... Lee S. Sobel

U.S.D.A. Forest Service - Forest Products Laboratory (G) .... Richard Bergman
Michael A. Ritter (Alt.)

Verdatek Solutions LLC (U) .... Matthew Belcher

Vinyl Siding Institute, Inc. (P) .... Matthew Dobson
Jery Y. Huntley (Alt.)

Winchester Homes, Inc. (U) .... Randall K. Melvin

Window & Door Manufacturers Association (P) .... Jeff Inks

Acknowledgement

The development of the 2012 edition of the *National Green Building Standard*™ could only be accomplished by the hard work of not only the members of the committee, but also the many non-committee members, including industry representatives and others, who actively participated and contributed to the process through their work on various Task Groups. The committee recognizes their contributions, as well as those who participated in the public hearings who, although not on the committee, provided valuable input during this development.

INTEREST CATEGORIES

Membership by Interest Category	
General Interest (G):	12
Producer Interest (P):	13
User Interest (U):	16
TOTAL:	**41**

TABLE OF CONTENTS

CHAPTER 1 SCOPE AND ADMINISTRATION 1

101 GENERAL 1

    101.1 Title...1
    101.2 Scope 1
    101.3 Intent 1
    101.4 Referenced documents 1
    101.5 Appendices 1

102 CONFORMANCE 1

    102.1 Mandatory practices 1
    102.2 Conformance language..1
    102.3 Documentation 1
    102.4 Alternative compliance methods 1

103 ADMINISTRATION...2

    103.1 Administration 2

CHAPTER 2 DEFINITIONS 3

201 GENERAL 3

    201.1 Scope 3
    201.2 Interchangeability 3
    201.3 Terms defined in other documents 3
    201.4 Terms not defined 3

202 DEFINITIONS 3

CHAPTER 3 COMPLIANCE METHOD 11

301 GENERAL 11

    301.1 Environmental rating levels 11
    301.2 Awarding of points..11

302 GREEN SUBDIVISIONS 11

    302.1 Site design and development 11

303 GREEN BUILDINGS 11

    303.1 Green buildings 11

304 GREEN MULTI-UNIT BUILDINGS 12

    304.1 Multi-unit buildings 12

305 GREEN REMODELING 12

    305.1 Compliance 12
    305.2 Compliance options..12
    305.3 Whole-building rating criteria 13
    305.4 Criteria for remodeled functional areas of buildings 14

306 GREEN ACCESSORY STRUCTURES 14

    306.1 Applicability 14
    306.2 Compliance 14

CHAPTER 4 SITE DESIGN AND DEVELOPMENT **15**

400 SITE DESIGN AND DEVELOPMENT 15

    400.0 Intent 15

401 SITE SELECTION 15

    401.0 Intent 15
    401.1 Infill site 15
    401.2 Greyfield site 15
    401.3 Brownfield site 15
    401.4 Low-slope site 15

402 PROJECT TEAM, MISSION STATEMENT, AND GOALS 15

    402.0 Intent 15
    402.1 Team 15
    402.2 Training 15
    402.3 Project checklist 15
    402.4 Development agreements 15

403 SITE DESIGN 16

    403.0 Intent 16
    403.1 Natural resources 16
    403.2 Building orientation 16
    403.3 Slope disturbance 16
    403.4 Soil disturbance and erosion 16
    403.5 Stormwater management 17
    403.6 Landscape plan 17
    403.7 Wildlife habitat 18
    403.8 Operation and maintenance plan 18
    403.9 Existing buildings 18
    403.10 Existing and recycled materials 18
    403.11 Environmentally sensitive areas 18

404 SITE DEVELOPMENT AND CONSTRUCTION 19

    404.0 Intent 19
    404.1 On-site supervision and coordination 19
    404.2 Trees and vegetation 19
    404.3 Soil disturbance and erosion 19
    404.4 Wildlife habitat 19

405 INNOVATIVE PRACTICES 20

    405.0 Intent 20
    405.1 Driveways and parking areas 20
    405.2 Street widths 20
    405.3 Cluster development 20
    405.4 Zoning 21
    405.5 Wetlands 21
    405.6 Multi-modal transportation 21
    405.7 Density 21
    405.8 Mixed-use development 21
    405.9 Open space 22
    405.10 Community garden(s) 22

CHAPTER 5 LOT DESIGN, PREPARATION, AND DEVELOPMENT **23**

500 LOT DESIGN, PREPARATION, AND DEVELOPMENT 23

    500.0 Intent 23

501 LOT SELECTION 23

    501.1 Lot 23
    501.2 Multi-modal transportation 23

502 PROJECT TEAM, MISSION STATEMENT, AND GOALS 23

    502.1 Project team, mission statement, and goals 23

503 LOT DESIGN 24

    503.0 Intent 24
    503.1 Natural resources 24
    503.2 Slope disturbance 24
    503.3 Soil disturbance and erosion 25
    503.4 Stormwater management 25
    503.5 Landscape plan 26
    503.6 Wildlife habitat 26
    503.7 Environmentally sensitive areas 26

504 LOT CONSTRUCTION 27

    504.0 Intent 27
    504.1 On-site supervision and coordination 27
    504.2 Trees and vegetation 27
    504.3 Soil disturbance and erosion implementation 27

505 INNOVATIVE PRACTICES 28

    505.0 Intent 28
    505.1 Driveways and parking areas 28
    505.2 Heat island mitigation 28
    505.3 Density 29
    505.4 Mixed-use development 29
    505.5 Community garden(s) 29

CHAPTER 6 RESOURCE EFFICIENCY **31**

601 QUALITY OF CONSTRUCTION MATERIALS AND WASTE 31

    601.0 Intent 31
    601.1 Conditioned floor area 31
    601.2 Material usage 31
    601.3 Building dimensions and layouts 31
    601.4 Framing and structural plans 32
    601.5 Prefabricated components 32
    601.6 Stacked stories 32
    601.7 Site-applied finishing materials 32
    601.8 Foundations 33
    601.9 Above grade wall systems 33

602 ENHANCED DURABILITY AND REDUCED MAINTENANCE 33

    602.0 Intent 33
    602.1 Moisture Management – Building Envelope 33
    602.2 Roof surfaces 36

602.3 Roof water discharge 36
602.4 Finished grade 36

603 REUSED OR SALVAGED MATERIALS 37

603.0 Intent 37
603.1 Reuse of existing building 37
603.2 Salvaged materials 37
603.3 Scrap materials 37

604 RECYCLED-CONTENT BUILDING MATERIALS 37

604.1 Recycled content 37

605 RECYCLED CONSTRUCTION WASTE 38

605.0 Intent 38
605.1 Construction waste management plan 38
605.2 On-site recycling 38
605.3 Recycled construction materials 38

606 RENEWABLE MATERIALS 38

606.0 Intent 38
606.1 Biobased products 38
606.2 Wood-based products 39
606.3 Manufacturing energy 39

607 RECYCLING AND WASTE REDUCTION 39

607.1 Recycling 39
607.2 Food waste disposers 39

608 RESOURCE-EFFICIENT MATERIALS 40

608.1 Resource-efficient materials 40

609 REGIONAL MATERIALS 40

609.1 Regional materials 40

610 LIFE CYCLE ANALYSIS 40

610.1 Life cycle analysis 40

611 INNOVATIVE PRACTICES 42

611.1 Manufacturer's environmental management system concepts 42
611.2 Sustainable products 42
611.3 Universal design elements 42

CHAPTER 7 ENERGY EFFICIENCY 47

701 MINIMUM ENERGY EFFICIENCY REQUIREMENTS 47

701.1 Mandatory requirements 47
701.1.1 Minimum Performance Path requirements 47
701.1.2 Minimum Prescriptive Path requirements 47
701.1.3 Alternative bronze level compliance 47
701.2 Emerald level points 47
701.3 Adopting Entity review 47
701.4 Mandatory practices 47
701.4.1 HVAC systems 47
701.4.2 Duct systems 48

701.4.3 Insulation and air sealing 48
701.4.4 High-efficacy lighting 50
701.4.5 Boiler supply piping 50

702 PERFORMANCE PATH 50

702.1 Point allocation 50
702.2 Energy cost performance levels 50

703 PRESCRIPTIVE PATH 51

703.1 Building envelope 51
703.2 HVAC equipment efficiency 55
703.3 Duct Systems 58
703.4 Water heating system 59
703.5 Lighting and appliances 61
703.6 Passive solar design 61

704 ADDITIONAL PRACTICES 64

704.1 Application of additional practice points 64
704.2 Lighting 64
704.3 Return ducts and transfer grilles 64
704.4 HVAC design and installation 64
704.5 Installation and performance verification 65

705 INNOVATIVE PRACTICES 66

705.1 Energy consumption control 66
705.2 Renewable energy service plan 66
705.3 Smart Appliances and Systems 66
705.4 Pumps 67
705.5 Additional renewable energy options 67
705.6 Parking garage efficiency 67

CHAPTER 8 WATER EFFICIENCY 69

801 INDOOR AND OUTDOOR WATER USE 69

801.0 Intent 69
801.1 Indoor hot water usage 69
801.2 Water-conserving appliances 71
801.3 Showerheads 71
801.4 Lavatory faucets 71
801.5 Water closets and urinals 72
801.6 Irrigation systems 72
801.7 Rainwater collection and distribution 73
801.8 Sediment filters 73

802 INNOVATIVE PRACTICES 74

802.1 Reclaimed, gray, or recycled water..74
802.2 Automatic shutoff water devices 74
802.3 Engineered biological system or intensive bioremediation system 74
802.4 Recirculating humidifier..74
802.5 Advanced wastewater treatment system 74

CHAPTER 9 INDOOR ENVIRONMENTAL QUALITY **75**

901 POLLUTANT SOURCE CONTROL 75

901.0 Intent 75
901.1 Space and water heating options 75
901.2 Solid fuel-burning appliances 76
901.3 Garages 76
901.4 Wood materials 76
901.5 Cabinets 77
901.6 Carpets 77
901.7 Hard-surface flooring 77
901.8 Wall coverings 78
901.9 Interior architectural coatings 78
901.10 Interior adhesives and sealants 80
901.11 Insulation 81
901.12 Carbon monoxide (CO) alarms 81
901.13 Building entrance pollutants control 81
901.14 Non-smoking areas 82

902 POLLUTANT CONTROL 82

902.0 Intent 82
902.1 Spot ventilation 82
902.2 Building ventilation systems 83
902.3 Radon control 83
902.4 HVAC system protection 83
902.5 Central vacuum systems 83
902.6 Living space contaminants 83

903 MOISTURE MANAGEMENT: VAPOR, RAINWATER, PLUMBING, HVAC 84

903.0 Intent 84
903.1 Plumbing 84
903.2 Duct insulation 84
903.3 Relative humidity 84

904 INNOVATIVE PRACTICES 84

904.1 Humidity monitoring system 84
904.2 Kitchen exhaust 84
Fig. 9(1) EPA Map of Radon Zones 85

CHAPTER 10 OPERATION, MAINTENANCE, AND BUILDING OWNER EDUCATION **87**

1001 BUILDING OWNERS' MANUAL FOR ONE- AND TWO-FAMILY DWELLINGS 87

1001.0 Intent 87

1002 TRAINING OF BUILDING OWNERS ON OPERATION AND MAINTENANCE FOR ONE- AND TWO-FAMILY DWELLINGS AND MULTI-UNIT BUILDINGS 88

1002.1 Training of building owners 88

1003 CONSTRUCTION, OPERATION, AND MAINTENANCE MANUALS AND TRAINING FOR MULTI-UNIT BUILDINGS 88

1003.0 Intent 88
1003.1 Building construction manual 89
1003.2 Operations manual 89

1003.3 Maintenance manual 90

1004 INNOVATIVE PRACTICES 90

1004.1 (Reserved) 90

CHAPTER 11 REMODELING **91**

11.500 LOT DESIGN, PREPARATION, AND DEVELOPMENT 91

11.500.0 Intent 91

11.501 LOT SELECTION 91

11.501.2 Multi-modal transportation 91

11.502 PROJECT TEAM, MISSION STATEMENT, AND GOALS 91

11.502.1 Project team, mission statement, and goals 91

11.503 LOT DESIGN 91

11.503.0 Intent 91

11.503.1 Natural resources 92

11.503.2 Slope disturbance 92

11.503.3 Soil disturbance and erosion 92

11.503.4 Storm water management 93

11.503.5 Landscape plan 93

11.503.6 Wildlife habitat 94

11.503.7 Environmentally sensitive areas 94

11.504 LOT CONSTRUCTION 94

11.504.0 Intent 94

11.504.1 On-site supervision and coordination 94

11.504.2 Trees and vegetation 94

11.504.3 Soil disturbance and erosion implementation 95

11.505 INNOVATIVE PRACTICES 95

11.505.0 Intent 95

11.505.1 Driveways and parking areas 95

11.505.2 Heat island mitigation 96

11.505.3 Density 96

11.505.4 Mixed-use development 96

11.505.5 Community Garden(s) 96

11.601 QUALITY OF CONSTRUCTION MATERIALS AND WASTE 97

11.601.0 Intent 97

11.601.1 Conditioned floor area 97

11.601.2 Material usage 97

11.601.3 Building dimensions and layouts 97

11.601.4 Framing and structural plans 98

11.601.5 Prefabricated components 98

11.601.6 Stacked stories 98

11.601.7 Site-applied finishing materials 98

11.601.8 Foundations 99

11.602 ENHANCED DURABILITY AND REDUCED MAINTENANCE 99

11.602.0 Intent 99

11.602.1 Moisture Management – Building Envelope 99

11.602.2 Roof surfaces 103
11.602.3 Roof water discharge 103
11.602.4 Finished grade 103

11.603 REUSED OR SALVAGED MATERIALS 103

11.603.0 Intent 103
11.603.1 Reuse of existing building 103
11.603.2 Salvaged materials 104
11.603.3 Scrap materials 104

11.604 RECYCLED-CONTENT BUILDING MATERIALS 104

11.604.1 Recycled content 104

11.605 RECYCLED CONSTRUCTION WASTE 104

11.605.0 Intent 104
11.605.1 Hazardous Waste 104
11.605.2 Construction waste management plan 104
11.605.3 On-site recycling 105
11.605.4 Recycled construction materials 105

11.606 RENEWABLE MATERIALS 105

11.606.0 Intent 105
11.606.1 Biobased products 105
11.606.2 Wood-based products 106
11.606.3 Manufacturing energy 106

11.607 RECYCLING AND WASTE REDUCTION 106

11.607.1 Recycling 106
11.607.2 Food waste disposers 106

11.608 RESOURCE-EFFICIENT MATERIALS 106

11.608.1 Resource-efficient materials 106

11.609 REGIONAL MATERIALS 107

11.609.1 Regional materials 107

11.610 LIFE CYCLE ANALYSIS 107

11.610.1 Life cycle analysis 107
11.610.1.1 Whole-building life cycle analysis 107
11.610.1.2 Life cycle analysis for a product or assembly 107

11.611 INNOVATIVE PRACTICES 108

11.611.1 Manufacturer's environmental management system concepts 108
11.611.2 Sustainable products 109
11.611.3 Universal design elements 109

11.701 MINIMUM ENERGY EFFICIENCY REQUIREMENTS 109

11.701.4 Mandatory practices 109
11.701.4.1 HVAC systems 109
11.701.4.2 Duct systems 110
11.701.4.3 Insulation and air sealing 110
11.701.4.4 High-efficacy lighting 112
11.701.4.5 Boiler supply piping 112

11.901 POLLUTANT SOURCE CONTROL 112

    11.901.0 Intent 112
    11.901.1 Space and water heating options..112
    11.901.2 Solid fuel-burning appliances 113
    11.901.3 Garages 114
    11.901.4 Wood materials 114
    11.901.5 Cabinets 114
    11.901.6 Carpets...115
    11.901.7 Hard-surface flooring 115
    11.901.8 Wall coverings...115
    11.901.9 Interior architectural coatings 115
    11.901.10 Adhesives and sealants 118
    11.901.11 Insulation 119
    11.901.12 Carbon monoxide (CO) alarms 119
    11.901.13 Building entrance pollutants control 119
    11.901.14 Non-smoking areas 119
    11.901.15 Lead-safe work practices 119

11.902 POLLUTANT CONTROL...119

    11.902.0 Intent 119
    11.902.1 Spot ventilation..119
    11.902.2 Building ventilation systems 120
    11.902.3 Radon control 121
    11.902.4 HVAC system protection..121
    11.902.5 Central vacuum systems..121
    11.902.6 Living space contaminants...121

11.903 MOISTURE MANAGEMENT: VAPOR, RAINWATER, PLUMBING, HVAC 121

    11.903.0 Intent 121
    11.903.1 Plumbing 121
    11.903.2 Duct insulation...121
    11.903.3 Relative humidity 122

11.904 INNOVATIVE PRACTICES 122

    11.904.1 Humidity monitoring system 122
    11.904.2 Kitchen exhaust..122

11.1001 BUILDING OWNERS' MANUAL FOR ONE- AND TWO-FAMILY DWELLINGS...........................122

    11.1001.0 Intent 122

11.1002 TRAINING OF BUILDING OWNERS ON OPERATION AND MAINTENANCE FOR ONE- AND TWO-
    FAMILY DWELLINGS AND MULTI-UNIT BUILDINGS 123

    11.1002.1 Training of building owners 123

11.1003 CONSTRUCTION, OPERATION, AND MAINTENANCE MANUALS AND TRAINING FOR MULTI-UNIT
    BUILDINGS 124

    11.1003.0 Intent 124
    11.1003.1 Building construction manual 124
    11.1003.2 Operations manual...124
    11.1003.3 Maintenance manual..125

11.1004 INNOVATIVE PRACTICES 126

CHAPTER 12 REMODELING OF FUNCTIONAL AREAS **127**

12.00 REMODELING OF FUNCTIONAL AREAS 127

    12.0 Intent.. 127
    12.0.1 Applicability 127

12.1 GENERAL 127

    12.1.601.2 Material usage 127
    12.1.602.1.7.1 Moisture control measures 127
    12.1.602.1.7.2 Moisture content 127
    12.1.602.1.11 Tile backing materials.. 127
    12.1(A) Product or material selection 128
    12.1(A).601.7 Site-applied finishing materials 128
    12.1(A).603.1 Reused and salvaged materials 128
    12.1(A).604.1 Recycled content 128
    12.1(A).606.1 Biobased products 128
    12.1(A).606.2 Wood-based products.. 128
    12.1(A).608.1 Resource-efficient materials 129
    12.1(A).609.1 Regional materials 129
    12.1(A).610.1 Life cycle analysis 129
    12.1(A).610.1.1 Functional area life cycle analysis 129
    12.1(A).610.1.2 Life cycle analysis for a product or assembly.. 129
    12.1(A).611.1 Manufacturer's environmental management system concepts 130
    12.1(A).611.2 Sustainable products 130
    12.1.605.0 Hazardous materials and waste 131
    12.1.701.4.1.1 HVAC system sizing 131
    12.1.701.4.2.1 Duct air sealing 131
    12.1.701.4.2.2 Supply ducts 131
    12.1.701.4.2.3 Duct system sizing 131
    12.1.701.4.3.1 Building thermal envelope 131
    12.1.701.4.3.2 Air sealing and insulation 131
    12.1.701.4.3.3 Fenestration air leakage 133
    12.1.701.4.3.4 Recessed lighting 133
    12.1.701.4.4 High-efficacy lighting 133
    12.1.701.4.5 Boiler supply piping 133
    12.1.703.5.3 Appliances 133
    12.1.901.1.4 Gas-fired equipment 133
    12.1.901.2.1 Solid fuel-burning appliances 133
    12.1.901.3 Attached garages 133
    12.1.901.4 Wood materials 134
    12.1.901.5 Cabinets 134
    12.1.901.6 Carpets 134
    12.1.901.7 Hard-surface flooring 134
    12.1.901.8 Interior wall coverings... 135
    12.1.901.9 Architectural coatings 135
    12.1.901.10 Adhesives and sealants 137
    12.1.901.11 Insulation 138
    12.1.901.15 Lead-safe work practices 138
    12.1.902.1.1 Spot ventilation 138
    12.1.902.4 HVAC system protection 138
    12.1.903.2 Duct insulation 138

12.2 KITCHEN REMODELS 139

    12.2.0 Applicability 139

    12.2.607.1 Recycling...139

    12.2.611.4 Food waste disposers 139

12.3 BATHROOM REMODELS 139

    12.3.0 Applicability 139

    12.3.611.3 Universal design elements...139

    12.3.801.4 Showerheads...139

    12.3.801.5.1 Faucets 139

    12.3.801.6 Water closets 139

12.4 BASEMENT REMODELS 140

    12.4.0 Applicability 140

    12.4.1 Moisture inspection 140

    12.4.2 Kitchen 140

    12.4.3 Bathroom...140

    12.4.902.3 Radon control 140

12.5 ADDITIONS 140

    12.5.0 Applicability 140

    12.5.1 Kitchen 140

    12.5.2 Bathroom...140

    12.5.503.5 Landscape plan..140

    12.5.602.1.1.1 Capillary break 140

    12.5.602.1.3.1 Exterior drain tile 140

    12.5.602.1.4.1 Crawlspace 141

    12.5.602.1.8 Water-resistive barrier 141

    12.5.602.1.9 Flashing...141

    12.5.602.1.14 Ice barrier 141

    12.5.602.1.15 Architectural features 141

    12.5.602.4.1 Finished grade 142

    12.5.902.3 Radon control..142

CHAPTER 13 REFERENCED DOCUMENTS...143

1301 GENERAL 143

1302 REFERENCED DOCUMENTS 143

APPENDIX A DUCTED GARAGE EXHAUST FAN SIZING CRITERIA.......................................155

APPENDIX B WHOLE BUILDING VENTILATION SYSTEM SPECIFICATIONS 157

APPENDIX C CLIMATE ZONES 161

APPENDIX D EXAMPLES OF THIRD-PARTY PROGRAMS FOR INDOOR ENVIRONMENTAL QUALITY 175

APPENDIX E ACCESSORY STRUCTURES 177

THIS PAGE INTENTIONALLY LEFT BLANK

CHAPTER 1

SCOPE AND ADMINISTRATION

101 GENERAL

101.1 Title. The title of this document is the *National Green Building Standard™*, hereinafter referred to as "this Standard."

101.2 Scope. The provisions of this Standard shall apply to design and construction of the residential portion(s) of any building, not classified as an institutional use, in all climate zones. This Standard shall also apply to subdivisions, building sites, building lots, accessory structures, and the residential portions of alterations, additions, renovations, mixed-use buildings, and historic buildings.

101.3 Intent. The purpose of this Standard is to establish criteria for rating the environmental impact of design and construction practices to achieve conformance with specified performance levels for green residential buildings, renovation thereof, accessory structures, building sites, and subdivisions. This Standard is intended to provide flexibility to permit the use of innovative approaches and techniques. This Standard is not intended to abridge safety, health, or environmental requirements contained in other applicable laws, codes, or ordinances.

101.4 Referenced documents. The codes, standards, and other documents referenced in this Standard shall be considered part of the requirements of this Standard to the prescribed extent of each such reference. The edition of the code, standard, or other referenced document shall be the edition referenced in Chapter 13.

101.5 Appendices. Where specifically required by a provision in this Standard, that appendix shall apply. Appendices not specifically required by a provision of this Standard shall not apply unless specifically adopted.

102 CONFORMANCE

102.1 Mandatory practices. This Standard does not require compliance with any specific practice except those noted as mandatory.

102.2 Conformance language. The green building provisions are written in mandatory language by way of using the verbs "to be," "is," "are," etc. The intent of the language is to require the user to conform to a particular practice in order to qualify for the number of points assigned to that practice. Where the term "shall" is used, or the points are designated as "mandatory," the provision or practice is mandatory.

102.3 Documentation. Verification of conformance to green building practices shall be the appropriate construction documents, architectural plans, site plans, specifications, builder certification and sign-off, inspection reports, or other data that demonstrates conformance as determined by the Adopting Entity. Where specific documentation is required by a provision of the Standard, that documentation is noted with that provision.

102.4 Alternative compliance methods. Alternative compliance methods shall be acceptable where the Adopting Entity finds that the proposed green building practice meets the intent of this Standard.

SECTION 103 ADMINISTRATION

103.1 Administration. The Adopting Entity shall specify performance level(s) to be achieved as identified in Chapter 3 and shall provide a verification process to ensure compliance with this Standard.

CHAPTER 2

DEFINITIONS

201 GENERAL

201.1 Scope. Unless otherwise expressly stated, the following words and terms shall, for the purposes of this Standard, have the meanings shown in this chapter.

201.2 Interchangeability. Words used in the present tense include the future; words stated in the masculine gender include the feminine and neuter; the singular number includes the plural and the plural, the singular.

201.3 Terms defined in other documents. Where terms are not defined in this Standard, and such terms are used in relation to the reference of another document, those terms shall have the definition in that document.

201.4 Terms not defined. Where terms are not defined through the methods authorized by this section, such terms shall have ordinarily accepted meanings such as the context implies.

SECTION 202 DEFINITIONS

ACCESSORY STRUCTURE. A structure, the use of which is customarily accessory to and incidental to that of the residential building; the structure is located on the same lot or site as the residential building; the structure does not contain a dwelling unit; and (1) is classified as Group U – Utility and Miscellaneous in accordance with the ICC International Building Code, or (2) is classified as accessory in accordance with the ICC International Residential Code, or (3) is classified as accessory to the residential use by a determination of the Adopting Entity.

ADDITION. An extension or increase in floor area or height of a building or structure.

ADOPTING ENTITY. The governmental jurisdiction, green building program, or any other third-party compliance assurance body that adopts this Standard, and is responsible for implementation and administration of the practices herein.

ADVANCED FRAMING. Code compliant layout, framing and engineering techniques that minimize the amount of framing products used and waste generated to construct a building while maintaining the structural integrity of the building.

AFUE (Annual Fuel Utilization Efficiency). The ratio of annual output energy to annual input energy which includes any non-heating season pilot input loss, and for gas or oil-fired furnaces or boilers, does not include electrical energy.

AIR BARRIER. Material(s) assembled and joined together to provide a barrier to air leakage through the building envelope. An air barrier may be a single material or a combination of materials.

AIR HANDLER. A blower or fan used for the purpose of distributing supply air to a room, space, or area.

AIR INFILTRATION. The uncontrolled inward air leakage into a building caused by the pressure effects of wind or the effect of differences in the indoor and outdoor air density or both.

AIR, MAKE-UP. Air that is provided to replace air being exhausted.

ARCHITECTURAL COATINGS. A material applied onto or impregnated into a substrate for protective, decorative, or functional purposes. Such materials include, but are not limited to, primers, paints, varnishes, sealers, and stains. An architectural coating is a material applied to stationary structures or their appurtenances at the site of installation. Coatings applied in shop applications, sealants, and adhesives are not considered architectural coatings.

BIOBASED PRODUCT. A commercial or industrial product used in site development or building construction that is composed, in whole or in significant part, of biological products, renewable agricultural materials (including plant, animal, and marine materials), or forestry materials.

BROWNFIELD (also EPA-Recognized Brownfield). Real property, the expansion, redevelopment, or reuse that may be complicated by the presence or potential presence of a hazardous substance, pollutant, or contaminant, and includes Brownfield Site as defined in Public Law 107-118 (H.R. 2869) - "Small Business Liability Relief and Brownfields Revitalization Act."

(i.e.: Pub.L. 107-118, § 1, Jan. 11, 2002, 115 Stat. 2356, provided that: "This Act [enacting 42 U.S.C.A. § 9628, amending this section, 42 U.S.C.A. § 9604, 42 U.S.C.A. § 9605, 42 U.S.C.A. § 9607, and 42 U.S.C.A. § 9622, and enacting provisions set out as notes under this section and 42 U.S.C.A. § 9607] may be cited as the 'Small Business Liability Relief and Brownfields Revitalization Act'.")

CERTIFIED GEOTHERMAL SERVICE CONTRACTOR. A person who has a current certification from the International Ground Source Heat Pump Association as an installer of ground source heat pump systems or as otherwise approved by the Adopting Entity.

CLIMATE ZONE. Climate zones are determined based on Figure 6(1).

CLUSTER DEVELOPMENT. A design technique that concentrates residential buildings and related infrastructure at a higher density within specified areas on a site. The remaining land on the site can then be used for low intensity uses such as recreation, common open space, farmland, or the preservation of historical sites and environmentally sensitive areas.

COGENERATION. An energy process that consecutively generates useful thermal and electric energy from the same fuel source.

COMMON AREA(S).
1. Areas within a site or lot that are predominantly open spaces and consist of non-residential structures, landscaping, recreational facilities, roadways and walkways, which are owned and maintained by an incorporated or chartered entity such as a homeowner's association or governmental jurisdiction; or
2. Areas of a multi-unit building that are outside the boundaries of a dwelling unit and are shared among or serve the dwelling units; including, but not limited to, hallways, amenity and resident services areas, parking areas, property management offices, mechanical rooms, and laundry rooms.

COMPOST FACILITY. An outdoor bin or similar structure designed for the decomposition of organic material such as leaves, twigs, grass clippings, and vegetative food waste.

CONDITIONED SPACE. An area or room within a building being heated or cooled, containing uninsulated ducts, or with a fixed opening directly into an adjacent conditioned space.

CONSTRUCTED WETLAND. An artificial wetland system (such as a marsh or swamp) created as new and/or restored habitat for native wetland plant and wildlife communities as well as to provide and/or restore wetland functions to the area. Constructed wetlands are often created as compensatory mitigation for ecological disturbances that result in a loss of natural wetlands from (1) anthropogenic discharge for wastewater, stormwater runoff, or sewage treatment; (2) mines or refineries; or (3) the development.

CONSTRUCTION WASTE MANAGEMENT PLAN. A system of measures designed to reduce, reuse, and recycle the waste generated during construction and to properly dispose of the remaining waste.

CONTINUOUS PHYSICAL FOUNDATION TERMITE BARRIER. An uninterrupted, non-chemical method of preventing ground termite infestation (e.g., aggregate barriers, stainless steel mesh, flashing, or plastic barriers).

COP (Coefficient of Performance). A measure of the heating efficiency of ground and air-source heat pumps defined as the ratio of the rate of heat provided by the heat pump to the rate of energy input, in consistent units, for a complete heat pump under defined operating conditions. (See EER as a measure of the cooling efficiency of heat pumps.)

DEMAND CONTROLLED HOT WATER LOOP. A hot water circulation (supply and return) loop with a pump that runs "on demand" when triggered by a user-activated switch or motion-activated sensor.

DESUPERHEATER. An auxiliary heat exchanger that uses superheated gases from an air conditioner's or heat pump's vapor-compression cycle to heat water.

DIRECT-VENT APPLIANCE. A fuel-burning appliance with a sealed combustion system that draws all air for combustion from the outside atmosphere and discharges all flue gases to the outside atmosphere.

DRAIN-WATER HEAT RECOVERY. A system to recapture the heat energy in drain water and use it to preheat cold water entering the water heater or other water fixtures.

DURABILITY. The ability of a building or any of its components to perform its required functions in its service environment over a period of time without unforeseen cost for maintenance or repair.

DWELLING UNIT. A single unit providing complete, independent living facilities for one or more persons, including permanent provisions for living, sleeping, eating, cooking, and sanitation.

EER (Energy Efficiency Ratio). A measure of the instantaneous energy efficiency of electric air conditioning defined as the ratio of net equipment cooling capacity in Btu/h to total rate of electric input in watts under designated operating conditions. When consistent units are used, this ratio becomes equal to COP. (See also Coefficient of Performance.)

ENERGY MANAGEMENT CONTROL SYSTEM. An integrated computerized control system that is intended to operate the heating, cooling, ventilation, lighting, water heating, and/or other energy-consuming appliances and/or devices for a building in order to reduce energy consumption. Also known as Building Automation Control (BAC) or Building Management Control System (BMCS).

ENERGY MONITORING DEVICE. A device installed within a building or dwelling unit that can provide near real-time data on whole building or dwelling unit energy consumption.

ENGINEERED WOOD PRODUCTS. Products that are made by combining wood strand, veneers, lumber or other wood fiber with adhesive or connectors to make a larger composite structure.

ENVIRONMENTAL IMPACT. See **LCA (Life Cycle Analysis/Assessment)**.

ENVIRONMENTALLY SENSITIVE AREAS.
1. Areas within wetlands as defined by federal, state, or local regulations;
2. Areas of steep slopes;
3. "Prime Farmland" as defined by the U.S. Department of Agriculture;
4. Areas of "critical habitat" for any federal or state threatened or endangered species;
5. Areas defined by state or local jurisdiction as environmentally sensitive.

EROSION CONTROLS. Measures that prevent soil from being removed by wind, water, ice, or other disturbance.

EXISTING BUILDING. Building completed and occupied prior to any renovation considered under this Standard.

EXISTING SUBDIVISION. An area of land, defined as "Site" in this Chapter, that has received all development approvals and has been platted and all infrastructure is complete at time of application to this Standard.

FROST-PROTECTED SHALLOW FOUNDATION. A foundation that does not extend below the design frost depth and is protected against the effects of frost in compliance with SEI/ASCE 32-01 or the provisions for frost-protected shallow foundations of the ICC IBC or IRC, as applicable.

GRADE PLANE. A reference plane representing the average of the finished ground level adjoining the building at all exterior walls. Where the finished ground level slopes away from the exterior walls, the reference plane shall be established by the lowest points within the area between the building and the lot line or, where the lot line is more than 6 feet (1830 mm) from the building, between the structure and a point 6 feet (1830 mm) from the building.

GRAY WATER. Waste discharged from lavatories, bathtubs, showers, clothes washers, and laundry trays.

GREYFIELD SITE. A previously developed site with abandoned or underutilized structures, and little or no contamination or perceived contamination.

GROUND SOURCE HEAT PUMP. Space conditioning and/or water heating systems that employ a geothermal resource such as the ground, groundwater, or surface water as both a heat source and a heat sink and use a reversible refrigeration cycle to provide both heating and cooling.

HARDSCAPE. Asphalt, concrete, masonry, stone, wood, and other non-plant elements external to the building shell on a landscape.

HEAT PUMP. An appliance having heating or heating/cooling capability and which uses refrigerants to extract heat from air, liquid, or other sources.

HIGH-EFFICACY LAMPS. Compact fluorescent lamps (CFL); light emitting diode (LED); T-8 or smaller diameter linear fluorescent lamps; or lamps with a minimum efficacy of 1) 60 lumens per watt for lamps over 40 watts, 2) 50 lumens per watt for lamps over 15 watts to 40 watts, or 3) 40 lumens per watt for lamps 15 watts or less.

HISTORIC BUILDINGS. Buildings that are listed in or eligible for listing in the National Register of Historic Places (NRHP) or designated as being of historic or architectural significance under an appropriate state or local law.

HSPF (Heating Seasonal Performance Factor). The total seasonal heating output of a heat pump, in Btu, divided by the total electric energy input during the same period, in watt-hours using a defined test methodology.

HYDROZONING. A landscape practice that groups plants with similar watering needs together in an effort to conserve water.

ICF (INSULATING CONCRETE FORMS). A concrete forming system using stay-in-place forms of rigid foam plastic insulation, a hybrid of cement and foam insulation, a hybrid of cement and wood chips, or other insulating material for constructing cast-in-place concrete walls.

IMPERVIOUS SURFACE. Hard-covered ground area that prevents/retards the entry of water into the soil at that location, resulting in water flowing to another location. (also see HARDSCAPE)

INDIRECT-FIRED WATER HEATER. A water storage tank, typically with no internal heating elements, that is connected by piping to an external heating source such as a gas or oil-fired boiler.

INFILL. A location including vacant or underutilized land that may apply to either a site or a lot and is located in an area served by existing infrastructure such as centralized water and sewer connections, roads, drainage, etc., and the site boundaries are adjacent to existing development on at least one side.

INTEGRATED PEST MANAGEMENT. A sustainable approach to managing pests by combining biological, cultural, physical, and chemical tools in a way that minimizes economic, health, and environmental risks.

LANDSCAPE PRACTICE (LANDSCAPING). Any activity that modifies the visible features of an area of land. It may include:
1. Living elements, such as flora or fauna;
2. Natural elements such as terrain shape, elevation, or bodies of water;
3. Created or installed elements such as fences or other material objects;
4. Abstract elements such as the weather and lighting conditions.

LAVATORY FAUCET. A valve for dispensing hot and/or cold water to a basin used for washing hands and face, but not for food preparation.

LCA (Life Cycle Analysis/Assessment). An accounting and evaluation of the environmental aspects and potential impacts of materials, products, assemblies, or buildings throughout their life (from raw material acquisition through manufacturing, construction, use, operation, demolition, and disposal).

LOT. A single parcel of land generally containing one primary structure or use. Lot development, as defined by this Standard, may include multiple ownership (such as with a condominium building) or multiple uses (such as with a mixed-use building). A lot is predominately represented by a single-family dwelling unit, a multifamily structure, or a mixed-use building also containing offices and shops. Lots may be located in urban, suburban, and rural locations. A lot may be located within a site. (also see SITE)

LOW-IMPACT DEVELOPMENT. A storm water management approach that attempts to recreate the predevelopment hydrology of a site by using lot level topography and landscape to deter storm water runoff and promote soil infiltration and recharge.

LOW-VOC (PRODUCTS). Products or materials with volatile organic compound (VOC) emissions equal to or below the established thresholds as defined in the referenced VOC emissions requirements for each applicable section in this document. (also see VOC)

MAJOR COMPONENT.
1. All structural members and structural systems.
2. Building materials or systems that are typically applied as a part of over 50% of the surface area of the foundation, wall, floor, ceiling, or roof assemblies.

MANUFACTURED HOME CONSTRUCTION. Three-dimensional sections of the complete building or dwelling unit built in a factory in conformance with the HUD Manufactured Home Construction and Safety Standards (24 CFR, Part 3280) and transported to the jobsite to be joined together on a foundation.

MASS WALLS. Above-grade masonry or concrete walls having a mass greater than or equal to 30 pounds per square foot (146 kg/m^2), solid wood walls having a mass greater than or equal to 20 pounds per square foot (98 kg/m^2), and any other walls having a heat capacity greater than or equal to 6 Btu/ft$^2 \bullet$°F [266 J/(m$^2 \bullet$ K)] with a minimum of 50 percent of the required R-value on the exterior side of the wall's centerline.

MERV (Minimum Efficiency Reporting Value). The Minimum Efficiency Reporting Value for filters in accordance with criteria contained in ASHRAE 52.2.

MINOR COMPONENT. Building materials or systems that are not considered a major component. (Also see Major Component)

MIXED-USE BUILDING. A building that incorporates more than one use (e.g., residential, retail, commercial) in a single structure.

MIXED-USE DEVELOPMENT. A project that incorporates more than one use (e.g., residential, retail, commercial) on the same site.

MODULAR CONSTRUCTION. Three-dimensional sections of the complete building or dwelling unit built in a factory and transported to the jobsite to be joined together on a permanent foundation.

MULTI-UNIT BUILDING. A building containing multiple dwelling units and classified as R-2 under the ICC IBC.

NET DEVELOPABLE AREA. The land on which buildings may be constructed. Any land where buildings cannot be constructed due to environmental restrictors, or that is used for infrastructure or public purposes such as parks, schools, etc., is not considered net developable area.

NEW CONSTRUCTION. Construction of a new building or construction that completely replaces more than 75 percent of an existing building.

OPEN SPACE. An area of land or water that (1) remains in its natural state, (2) is used for agriculture, or (3) is free from intensive development.

PANELIZED ASSEMBLIES. Factory-assembled wall panels, roof trusses, and/or other components installed on-site.

PERFORMANCE PATH. An alternative set of standards (to the Prescriptive Path) with defined performance metrics, as specified in Chapter 7 of this Standard.

PERMEABLE MATERIAL. A material that permits the passage of water vapor and/or liquid.

PLUMBING FIXTURE. A receptor or device that requires both a water-supply connection and a discharge to the drainage system, such as water closets, lavatories, bathtubs, and sinks.

PRECUT. Materials cut to final size prior to delivery to site and ready for assembly.

PRESCRIPTIVE PATH. A set of provisions in a code or standard that must be adhered to for compliance.

PRESERVATION. The process of applying measures to maintain and sustain the existing materials, integrity, and/or form of a building, including its structure and building artifacts.

PROGRAMMABLE COMMUNICATING THERMOSTAT. A whole building or whole dwelling unit thermostat that can be monitored and controlled remotely.

PROJECTION FACTOR. The ratio of the overhang width to the overhang height above the door threshold or window sill (PF = A/B).

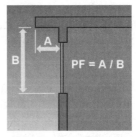

Projection Factor

R-VALUE (THERMAL RESISTANCE). The inverse of the time rate of heat flow through a body from one of its bounding surfaces to the other surface for a unit temperature difference between the two surfaces, under steady state conditions, per unit area ($h \times ft^2 \times °F/Btu$) [$(m^2 \times K)/W$].

RECYCLE. To recover and reprocess manufactured goods into new products.

RECYCLED CONTENT. Resources containing post-consumer or pre-consumer (post-industrial) recycled content.

> **POST-CONSUMER RECYCLED CONTENT.** Proportion of recycled material in a product generated by households or by commercial, industrial, and institutional facilities in their role as end users of the product that can no longer be used for its intended purpose. This includes returns of material from the distribution chain.

> **PRE-CONSUMER (POST-INDUSTRIAL) RECYCLED CONTENT.** Proportion of recycled material in a product diverted from the waste stream during the manufacturing process. Pre-consumer recycled

content does not include reutilization of materials such as rework, regrind, or scrap generated in a process and capable of being reclaimed within the same process that generated it.

REGIONAL MATERIAL. Material that originates, is produced, grows naturally, or occurs naturally within: (1) 500 miles (804.7 km) of the construction site if transported by truck or (2) 1,500 miles (2,414 km) of the construction site if transported for not less than 80 percent of the total transport distance by rail or water. Products that are assembled or produced from multiple raw materials are considered regional materials if the weighted average (by weight or volume) of the distance the raw materials have been transported meet the distance criteria.

REMODELING. The process of restoring or improving an existing building, dwelling unit, or property.

RENEWABLE ENERGY. Energy derived from sources that are regenerative or cannot be depleted.

RENEWABLE ENERGY SOURCE. Source of energy (excluding minerals) derived from incoming solar radiation, including natural solar radiation itself, photosynthetic processes; from phenomenon resulting therefrom, including wind, hydropower, waves and tides, and lake or pond thermal differences; from decomposition of waste material, including methane from landfills; from processes that use regenerated materials, including wood and bio-based products; and from the internal heat of the earth, including nocturnal thermal exchanges.

REPLACEMENT. The act or process of replacing material or systems.

REUSE. To recover a material or product for use again without reprocessing.

SEDIMENT CONTROLS. Practices used on building sites to minimize the movement of sand, soil, and particulates or dust from construction from reaching waterways.

SEER (Seasonal Energy Efficiency Ratio). The total cooling output of an electric air conditioner (or heat pump) during its normal annual usage period for cooling, in Btu, divided by the total electric energy input during the same period, in watt-hours (Wh), expressed as Btu/Wh. SEER is the cooling performance equivalent measurement of HSPF.

SHGC (Solar Heat Gain Coefficient). The ratio of the solar heat gain entering the space through the fenestration assembly to the incident solar radiation. Solar heat gain includes directly transmitted solar heat and absorbed solar radiation which is then reradiated, conducted, or convected into the space.

SIP (Structural Insulated Panel). A structural sandwich panel that consists of a light-weight foam plastic core securely laminated between two thin, rigid wood structural panel facings; a structural panel that consists of lightweight foam plastic and cold-formed steel sheet or structural cold-formed steel members; or other similar non-interrupted structural panels.

SITE. Any area of land that is or will be developed into two or more parcels of land intended for multiple ownership, uses, or structures and designed to be part of an integrated whole such as a residential subdivision, mixed-use development, or master-planned community. Site, as defined, generally contains multiple lots. (also see LOT)

SMART APPLIANCE. A product that has the capability to receive, interpret, and act on a signal transmitted by a utility, third-party energy service provider, or home energy management device, and automatically adjust its operation depending on both the signal's contents and settings by the consumer. The product has this capability either built-in or added through an external device that easily connects to the appliance.

SOLID FUEL-BURNING APPLIANCE. A chimney connected device designed for purposes of heating, cooking, or both that burns solid fuel.

STEEP SLOPES. Slopes equal to or greater than 25 percent (≥ 25%)

STORY. That portion of a building included between the upper surface of a floor and the upper surface of the floor or roof next above.

STORY ABOVE GRADE. Any story having its finished floor surface entirely above grade, except that a basement shall be considered as a story above grade where the finished surface of the floor above the basement is:
1. More than 6 feet (1829 mm) above grade plane.
2. More than 6 feet (1829 mm) above the finished ground level for more than 50 percent of the total building perimeter.
3. More than 12 feet (3658 mm) above the finished ground level at any point.

SUBDIVISION. A tract, lot, or parcel of land divided into two or more lots, plats, sites, or other divisions of land.

SWPPP (Stormwater Pollution Prevention Plan). A site-specific, written document or report that identifies required features specifically represented in the National Pollutant Discharge Elimination System (NPDES) Construction General Permit (CGP).

UA. The total U-factor times area for a component or building.

URBAN. Areas within a designated census tract of 1,000 people per square mile or located within a Metropolitan Statistical Area primary city, as designated by the U.S. Census Bureau.

U-FACTOR (THERMAL TRANSMITTANCE). The coefficient of heat transmission (air to air) through a building envelope component or assembly, equal to the time rate of heat flow per unit area and unit temperature difference between the warm side and cold side air films (Btu/h • ft^2 • °F) [W/(m^2 • K]).

VENTILATION. The natural or mechanical process of supplying conditioned or unconditioned air to, or removing such air from, any space.

VOC (Volatile Organic Compounds). A class of carbon-based molecules in substances and organic compounds that readily release gaseous vapors at room temperature as indoor pollutants and when reacting with other exterior pollutants can produce ground-level ozone.

WASTE HEAT. Heat discharged as a byproduct of one process to provide heat needed by a second process.

WATER FACTOR (WATER CONSUMPTION FACTOR). The quotient of the total weighted per-cycle water consumption divided by the capacity of the clothes washer.

WATER-RESISTIVE BARRIER. A material behind an exterior wall covering that is intended to resist liquid water that has penetrated behind the exterior covering from further intruding into the exterior wall assembly.

WETLANDS. Areas that are saturated by surface or ground water at a frequency and duration sufficient to support, and that under normal circumstances do support, a prevalence of vegetation typically adapted for life in saturated soil conditions. Wetlands are further defined by the EPA in the *Code of Federal Regulations*.

WILDLIFE HABITAT/CORRIDOR. An ecological or environmental area that is inhabited by a particular species of animal, plant, or other type of organism. It is the natural environment in which an organism lives or the physical environment that surrounds (influences and is utilized by) a species population.

WOOD-BASED PRODUCT. Any material that consists of a majority of wood or constituents derived from wood (e.g., wood fiber) as measured by either weight or volume.

CHAPTER 3

COMPLIANCE METHOD

301 GENERAL

301.1 Environmental rating levels. The building, project, site, and/or development environmental rating level shall consist of all mandatory requirements plus points assessed using the point system specified within this chapter. The rating level shall be in accordance with Section 302, 303, 304, or 305.3, as applicable. The designation for remodeled functional areas shall be in accordance with Section 305.4. The designation for accessory structures shall be in accordance with Section 306.

301.2 Awarding of points. Points shall be awarded as follows:

(1) The maximum number of points that can be awarded for each practice is noted with that practice.

(2) Point allocation for multi-unit buildings shall be as prescribed in Section 304.

(3) The Adopting Entity shall allow the use of new and innovative products and practices deemed to meet the intent of this Standard. Points assigned for any new product or practice shall be determined by the Adopting Entity. A maximum of 20 points may be awarded at the discretion of the Adopting Entity. Innovative practices and products shall fall under Chapters 5-10 (Categories 1-6 in Table 303); however these points shall only be assigned under Category 7. Point values shall be determined by comparing the innovative product or practice to a practice or product already described in the Standard. The applicant shall supply demonstrable, quantified data to support the innovative product or practice and to determine the practice's functional equivalent in the Standard for the points to be awarded.

302 GREEN SUBDIVISIONS

302.1 Site design and development. The threshold points required for the environmental rating levels to qualify a new or existing subdivision as green under this Standard shall be in accordance with Table 302 and based on points in Chapter 4.

Table 302
Threshold Point Ratings for Site Design and Development

Green Subdivision Category		Rating Level Points			
		One Star	Two Stars	Three Stars	Four Stars
Chapter 4	Site Design and Development	95	122	149	176

303 GREEN BUILDINGS

303.1 Green buildings. The threshold points required for the environmental rating levels for a green building shall be in accordance with Table 303. To qualify for one of these rating levels, all of the following shall be satisfied:

(1) The threshold number of points, in accordance with Table 303, shall be achieved as prescribed in Categories 1 through 6. The lowest level achieved in any category shall determine the overall rating level achieved for the building.

(2) In addition to the threshold number of points in each category, all mandatory provisions of each category shall be implemented.

(3) In addition to the threshold number of points prescribed in Categories 1 through 6, the additional points prescribed in Category 7 shall be achieved from any of the categories. Where deemed appropriate by the Adopting Entity based on regional conditions, additional points from Category 7 may be assigned to another category (or categories) to increase the threshold points required for that category (or categories). Points shall not be reduced by the Adopting Entity in any of the six other categories.

Table 303
Threshold Point Ratings for Green Buildings

Green Building Categories			Rating Level Points [1] [2]			
			BRONZE	SILVER	GOLD	EMERALD
1.	Chapter 5	Lot Design, Preparation, and Development	50	64	93	121
2.	Chapter 6	Resource Efficiency	43	59	89	119
3.	Chapter 7	Energy Efficiency	30	60	80	100
4.	Chapter 8	Water Efficiency	25	39	67	92
5.	Chapter 9	Indoor Environmental Quality	25	42	69	97
6.	Chapter 10	Operation, Maintenance, and Building Owner Education	8	10	11	12
7.		Additional Points from Any Category	50	75	100	100
		Total Points:	231	349	509	641

[1]   In addition to the threshold number of points in each category, all mandatory provisions of each category shall be implemented.

[2]   For dwelling units greater than 4,000 square feet (372 m^2), the number of points in Category 7 (Additional Points from Any Category) shall be increased in accordance with Section 601.1. The "Total Points" shall be increased by the same number of points.

304 GREEN MULTI-UNIT BUILDINGS

304.1 Multi-unit buildings. All residential portions of a building shall meet the requirements of this Standard. Partial compliance shall not be allowed. Unless otherwise noted, all units and residential common areas within a multi-unit building shall: 1) meet all mandatory requirements; and 2) achieve the point threshold required for the chosen environmental rating level in accordance with Table 303; and 3) achieve the same environmental rating level. Points for the green building practices that apply to multiple units shall be credited once for the entire building. Where points are credited, including where a weighted average is used, practices shall be implemented in all units, as applicable. Where application of a prescribed practice allows for a different number of points for different units in a multi-unit building, the fewer number of points shall be awarded, unless noted that a weighted average is used.

305 GREEN REMODELING

305.1 Compliance. Compliance with Section 305 shall be voluntary unless specifically adopted as mandatory by the Adopting Entity.

305.2 Compliance options. The criteria for existing buildings shall be in accordance with Section 305.3 for whole-building ratings or Section 305.4 for compliance designations of building functional areas.

305.3 Whole-building rating criteria

305.3.1 Applicability. The provisions of Section 305.3 shall apply to remodeling of existing buildings. In addition to the foundation, at least one major structural system (such as walls) of the existing building shall remain in place after the remodel for the building to be eligible for compliance under Section 305.3.

305.3.1.1 Additions. For a remodeled building that includes an addition, the entire building including the addition shall comply with the criteria of Section 305.3. The total above-grade conditioned area added during a remodel shall not exceed 75% of the existing building's above-grade conditioned area. For multi-unit buildings, the above-grade conditioned area shall be based on the entire building including all dwelling units and common areas.

305.3.2 Rating scope. The building rating achieved under Section 305.3 and the associated compliance criteria apply to the entire building after the remodel including any additions.

305.3.3 Mandatory practices. The building, including any additions and common areas, shall satisfy all practices designated as mandatory in Chapter 11.

305.3.4 Rating level. A minimum rating level of Bronze shall be achieved in each of the following categories: Energy efficiency (Sections 305.3.5), Water efficiency (Section 305.3.6), and Prescriptive practices (Section 305.3.7). The building rating level shall be the lowest rating level achieved in Sections 305.3.5, 305.3.6, and 305.3.7.

305.3.5 Energy efficiency. The energy efficiency rating level shall be based on the reduction in energy consumption resulting from the remodel in accordance with Table 305.3.5.

Table 305.3.5
Energy Rating Level Thresholds

	Rating Level			
	BRONZE	**SILVER**	**GOLD**	**EMERALD**
Reduction in energy consumption	15%	25%	35%	45%

305.3.5.1 Energy consumption reduction. The reduction in energy consumption resulting from the remodel shall be based on the estimated annual energy cost savings as determined by a third-party energy audit and analysis or utility consumption data. The reduction shall be the percentage difference between the consumption per square foot before and after the remodel calculated as follows:

[(consumption per square foot before remodel – consumption per square foot after remodel)/consumption per square foot before remodel]*100%

The occupancy and lifestyle assumed and the method of making the energy consumption estimates shall be the same for estimates before and after the remodel. The building configuration for the after-remodel estimate shall include any additions to the building or other changes to the configuration of the conditioned space. For multi-unit buildings, the energy consumption shall be based on the entire building including all dwelling units and common areas.

305.3.6 Water efficiency. The water efficiency rating level shall be based on the reduction in water consumption resulting from the remodel in accordance with Table 305.3.6.

Table 305.3.6
Water Rating Level Thresholds

	Rating Level			
	BRONZE	**SILVER**	**GOLD**	**EMERALD**
Reduction in water consumption	20%	30%	40%	50%

305.3.6.1 Water consumption reduction. Water consumption shall be based on the estimated annual use as determined by audit and analysis or use of utility consumption data. The reduction shall be the percentage difference between the consumption before and after the remodel calculated as follows:

[(consumption before remodel − consumption after remodel)/consumption before remodel]∗100%

The occupancy and lifestyle assumed and the method of making the water consumption estimates shall be the same for estimates before and after the remodel. The building configuration for the after-remodel estimate shall include any changes to the configuration of the building such as additions or new points of water use. For multi-unit buildings, the water consumption shall be based on the entire building including all dwelling units and common areas.

305.3.7 Prescriptive practices. The point thresholds for the environmental rating levels based on compliance with the Chapter 11 prescriptive practices shall be in accordance with Table 305.3.7. Any practice listed in Chapter 11 shall be eligible for contributing points to the prescriptive threshold ratings. The attributes of the existing building that were in compliance with the prescriptive practices of Chapter 11 prior to the remodel and remain in compliance after the remodel shall be eligible for contributing points to the prescriptive threshold ratings.

Table 305.3.7
Prescriptive Threshold Point Ratings

	Rating Level			
	BRONZE	SILVER	GOLD	EMERALD
Chapter 11 prescriptive thresholds	88	125	181	225

305.4 Criteria for remodeled functional areas of buildings

305.4.1 Applicability. The provisions of Section 305.4 shall apply to remodeling of one or more of the following functional areas of the existing building as follows:

1. Addition, kitchen, bathroom, or basement in buildings other than multi-unit buildings.
2. Kitchen or bathroom of an individual dwelling unit in a multi-unit building.

305.4.1.1 Additions. The total above-grade conditioned area added during a remodel shall not exceed 400 square feet.

305.4.2 Compliant. Projects that meet all applicable requirements of Chapter 12 for that functional area shall be designated as *compliant*.

305.4.3 Designation. The designation achieved under Section 305.4 applies only to the specific functional area of the existing building. The existing building may have more than one *compliant* functional area.

305.4.4 Additions. A bathroom(s), kitchen, or finished basement included in an addition shall comply with all criteria specifically applicable to those functional areas in accordance with the provisions of Chapter 12.

305.4.5 Mandatory. Projects shall satisfy all applicable practices designated as mandatory in Chapter 12.

305.4.6 Existing attributes. The attributes of the existing building that were in compliance with the applicable provisions of Chapter 12 prior to the remodel and remain in compliance after the remodel shall be eligible for contributing to demonstration of compliance under Section 305.4.

306 GREEN ACCESSORY STRUCTURES

306.1 Applicability. The designation criteria for accessory structures shall be in accordance with Appendix E.

306.2 Compliance. Compliance with Appendix E shall be voluntary unless specifically adopted as mandatory. If specifically adopted, the adopting entity shall establish rules for compliance with Appendix E.

CHAPTER 4

SITE DESIGN AND DEVELOPMENT

GREEN BUILDING PRACTICES	POINTS

400 SITE DESIGN AND DEVELOPMENT

400.0 Intent. This section applies to land development for the eventual construction of buildings or additions thereto that contain dwelling units. The rating earned under Section 302 based on practices herein, applies only to the site as defined in Chapter 2. The buildings on the site achieve a separate rating level or designation by complying with the provisions of Section 303, 304, 305, or 306, as applicable.

401 SITE SELECTION

401.0 Intent. The site is selected to minimize environmental impact by one or more of the following:

401.1 Infill site. An infill site is selected.	7
401.2 Greyfield site. A greyfield site is selected.	7
401.3 Brownfield site. A brownfield site is selected.	8
401.4 Low-slope site. A site with an average slope calculation of less than 15 percent is selected.	5

402 PROJECT TEAM, MISSION STATEMENT, AND GOALS

402.0 Intent. The site is designed and constructed by a team of qualified professionals trained in green development practices.

402.1 Team. A knowledgeable team is established and team member roles are identified with respect to green lot design, preparation, and development. The project's green goals and objectives are written into a mission statement.	4
402.2 Training. Training is provided to on-site supervisors and team members regarding the green development practices to be used on the project.	3
402.3 Project checklist. A checklist of green development practices to be used on the project is created, followed, and completed by the project team regarding the site.	**Mandatory** 4
402.4 Development agreements. Through a developer agreement or equivalent, the developer requires purchasers of lots to construct the buildings in compliance with this Standard (or equivalent) certified to a minimum bronze rating level.	6

GREEN BUILDING PRACTICES	POINTS

403
SITE DESIGN

403.0 Intent. The project is designed to avoid detrimental environmental impacts, minimize any unavoidable impacts, and mitigate for those impacts that do occur. The project is designed to minimize environmental impacts and to protect, restore, and enhance the natural features and environmental quality of the site.

(To acquire points allocated for the design, the intent of the design is implemented.)

403.1 Natural resources. Natural resources are conserved by one or more of the following:

(1)	A natural resources inventory is used to create the site plan.	**Mandatory 5**
(2)	A plan to protect and maintain priority natural resources/areas during construction is created. (also see Section 404 for guidance in forming the plan.)	**Mandatory 5**
(3)	Member of builder's project team participates in a natural resources conservation program.	4
(4)	Streets, buildings, and other built features are located to conserve high priority vegetation.	5

403.2 Building orientation. A minimum of 75 percent of the building sites are designed with the longer dimension of the structure to face within 20 degrees of south.	6

403.3 Slope disturbance. Slope disturbance is minimized by one or more of the following:

(1)	Hydrological/soil stability study is completed and used to guide the design of all buildings on the site.	5
(2)	All or a percentage of roads are aligned with natural topography to reduce cut and fill.	
	(a) 10 percent to 25 percent	1
	(b) 25 percent to 75 percent	4
	(c) greater than 75 percent	6
(3)	Long-term erosion effects are reduced by the use of clustering, terracing, retaining walls, landscaping, and restabilization techniques.	6

403.4 Soil disturbance and erosion. A site Stormwater Pollution Prevention Plan (SWPPP) is developed in accordance with applicable stormwater Construction General Permits. The plan includes one or more of the following:

(1)	Construction activities are scheduled to minimize length of time that soils are exposed.	4
(2)	Utilities are installed by alternate means such as directional boring in lieu of open-cut trenching. Shared easements or common utility trenches are utilized to minimize earth disturbance. Low ground pressure equipment or temporary matting is used to minimize excessive soil consolidation.	5
(3)	Limits of clearing and grading are demarcated.	4

GREEN BUILDING PRACTICES	POINTS
403.5 Stormwater management. Stormwater management design includes one or more of the following low-impact development techniques:	
(1) Natural water and drainage features are preserved and used.	7
(2) Vegetative swales, French drains, wetlands, drywells, rain gardens, and similar infiltration features are used.	6
(3) Permeable materials are selected/specified for common area roads, driveways, parking areas, walkways, and patios.	
(a) 10 percent to 25 percent	2
(b) 25 percent to 75 percent	5
(c) greater than 75 percent	8
(4) Stormwater management practices are selected/specified that manage rainfall on-site and prevent the off-site discharge from all storms up to and including the volume of the 95th percentile storm event.	7
(5) A hydrologic analysis is conducted that results in the design of a stormwater management system that maintains the pre-development (stable, natural) runoff hydrology of the site throughout the development or redevelopment process. Post-construction runoff rate, volume, and duration do not exceed predevelopment rates.	7
(6) Stormwater management features/structures are designed for the reduction of nitrogen, phosphorus, and sediment.	7

GREEN BUILDING PRACTICES	POINTS
403.6 Landscape plan. A landscape plan is developed to limit water and energy use in common areas while preserving or enhancing the natural environment utilizing one or more of the following:	
(1) A plan is formulated to restore or enhance natural vegetation that is cleared during construction. Landscaping is phased to coincide with achievement of final grades to ensure denuded areas are quickly vegetated.	6
(2) On-site native or regionally appropriate trees and shrubs are conserved, maintained, and reused for landscaping to the greatest extent possible.	6
(3) Turf grass species, other vegetation, and trees that are native or regionally appropriate for local growing conditions are selected.	5
(4) The percentage of all turf areas are limited as part of the landscaping.	
(a) 0 percent or EPA WaterSense Water Budget Tool is used to determine the maximum percentage of turf areas	6
(b) greater than 0 percent to less than 20 percent	5
(c) 20 percent to less than 40 percent	3
(d) 40 percent to 60 percent	2
(5) Plants with similar watering needs are grouped (hydrozoning).	4
(6) Species and locations for tree planting are identified and utilized to increase summer shading of streets, parking areas, and buildings and to moderate temperatures.	5
(7) Vegetative wind breaks or channels are designed as appropriate to local conditions.	4

GREEN BUILDING PRACTICES	POINTS
(8) On-site tree trimmings or stump grinding of regionally appropriate trees are used to provide protective mulch during construction or as base for walking trails, and cleared trees are recycled as sawn lumber or pulp wood.	4
(9) An integrated common area pest management plan to minimize chemical use in pesticides and fertilizers is developed.	4
(10) Plans for the common area landscape watering system include a weather-based or moisture-based controller. Required irrigation systems are designed in accordance with the Irrigation Association's *Turf and Landscape Best Management Practices*.	6
(11) Trees that might otherwise be lost due to site construction are transplanted to other areas on-site or off-site using tree-transplanting techniques to ensure a high rate of survival.	4
(12) Gray water irrigation systems are used to water common areas. Gray water used for irrigation conforms to all criteria of Section 802.1.	7
(13) Cisterns, rain barrels, and similar tanks are designed to intercept and store runoff. These systems may be above or below ground, and they may drain by gravity or be pumped. Stored water may be slowly released to a pervious area, and/or used for irrigation of lawn, trees, and gardens located in common areas.	6

403.7 Wildlife habitat. Measures are planned that will support wildlife habitat.	6

403.8 Operation and maintenance plan. An operation and maintenance plan (manual) is prepared and outlines ongoing service of common open area, utilities (storm water, waste water), and environmental management activities.	6

403.9 Existing buildings. Existing building(s) and structure(s) is/are preserved, reused, modified, or disassembled for reuse or recycling of building materials.	8

403.10 Existing and recycled materials. Existing or recycled materials are used as follows. **(Points awarded for every 10 percent of total construction and demolition materials that are reused, deconstructed, and/or salvaged. The percentage is consistently calculated on a weight, volume, or cost basis.)**	3
(1) Existing pavements, curbs, and aggregates are salvaged or reincorporated into the development.	
(2) Recycled asphalt or concrete is utilized in the project.	

403.11 Environmentally sensitive areas. Environmentally sensitive areas are as follows:		
(1) Environmentally sensitive areas including steep slopes, prime farmland, critical habitats, and wetlands are avoided as follows:		
	(a) <25 percent of site undeveloped	2
	(b) 25 percent – 75 percent of site undeveloped	4
	(c) >75 percent of site undeveloped	7
(2) Compromised environmentally sensitive areas are mitigated or restored.		4

ICC 700-2012 NATIONAL GREEN BUILDING STANDARD™

GREEN BUILDING PRACTICES	POINTS

404
SITE DEVELOPMENT AND CONSTRUCTION

404.0 Intent. Environmental impact during construction is avoided to the extent possible; impacts that do occur are minimized, and any significant impacts are mitigated.	

404.1 On-site supervision and coordination. On-site supervision and coordination is provided during clearing, grading, trenching, paving, and installation of utilities to ensure that specified green development practices are implemented. (also see Section 403.4)	5

404.2 Trees and vegetation. Designated trees and vegetation are preserved by one or more of the following:

(1)	Fencing or equivalent is installed to protect trees and other vegetation.	4
(2)	Trenching, significant changes in grade, compaction of soil, and other activities are avoided in critical root zones (canopy drip line) in "tree save" areas.	5
(3)	Damage to designated existing trees and vegetation is mitigated during construction through pruning, root pruning, fertilizing, and watering.	4

404.3 Soil disturbance and erosion. On-site soil disturbance and erosion are minimized by implementation of one or more of the following:

(1)	Limits of clearing and grading are staked out prior to construction.	5
(2)	"No disturbance" zones are created using fencing or flagging to protect vegetation and sensitive areas from construction vehicles, material storage, and washout.	4
(3)	Sediment and erosion controls are installed and maintained.	5
(4)	Topsoil is stockpiled and covered with tarps, straw, mulch, chipped wood, vegetative cover, or other means capable of protecting it from erosion for later use to establish landscape plantings.	5
(5)	Soil compaction from construction equipment is reduced by distributing the weight of the equipment over a larger area by laying lightweight geogrids, mulch, chipped wood, plywood, OSB (oriented strand board), metal plates, or other materials capable of weight distribution in the pathway of the equipment.	4
(6)	Disturbed areas are stabilized within the EPA-recommended 14-day period.	4
(7)	Soil is improved with organic amendments and mulch.	4

404.4 Wildlife habitat. Measures are implemented to support wildlife habitat.

(1)	Wildlife habitat is maintained.	5
(2)	Measures are instituted to establish or promote wildlife habitat.	5
(3)	Open space is preserved as part of a wildlife corridor.	6
(4)	Builder or member of builder's project team participates in a wildlife conservation program.	5

GREEN BUILDING PRACTICES	POINTS

405
INNOVATIVE PRACTICES

405.0 Intent. Innovative site design, preparation, and development practices are used to enhance environmental performance. Waivers or variances from local development regulations are obtained, and innovative zoning practices are used to implement such practices, as applicable.

405.1 Driveways and parking areas. Driveways and parking areas are minimized by one or more of the following:

(1)	Off-street parking areas are shared or driveways are shared; on-street parking is utilized; and alleys (shared common area driveways) are used for rear-loaded garages.	5
(2)	In multi-unit projects, parking capacity is not to exceed the local minimum requirements.	5
(3)	Structured parking is utilized to reduce the footprint of surface parking areas.	
	(a) 25 percent to less than 50 percent	3
	(b) 50 percent to 75 percent	5
	(c) greater than 75 percent	8

405.2 Street widths.

(1)	Street pavement widths are minimized per local code and are in accordance with Table 405.2.	6

Table 405.2
Maximum Street Widths

Facility Type	Maximum Width
Collector street with parking (one side only)	31 feet
Collector street without parking	26 feet
Local access with parking (one side only)	27 feet
Local access street without parking	20 feet
Queuing (one-lane) streets with parking	24 feet
Alleys and queuing (one-lane) streets without parking	17 feet

For SI: 1 foot = 304.8 mm

(2)	A waiver was secured by the developer from the local jurisdiction to allow for construction of streets below minimum width requirement.	8

405.3 Cluster development. Cluster development enables and encourages flexibility of design and development of land in such a manner as to preserve the natural and scenic qualities of the site by utilizing an alternative method for the layout, configuration and design of lots, buildings and structures, roads, utility lines and other infrastructure, parks, and landscaping.	10

     ICC 700-2012 NATIONAL GREEN BUILDING STANDARD™

GREEN BUILDING PRACTICES	POINTS

405.4 Zoning. Innovative zoning techniques are implemented in accordance with the following:		
(1)	Innovative zoning ordinances or local laws are used or developed for permissible adjustments to population density, area, height, open space, mixed-use, or other provisions for the specific purpose of open space, natural resource preservation or protection and/or mass transit usage. Other innovative zoning techniques may be considered on a case-by-case basis.	8
(2)	An increase to the permissible density, area, height, use, or other provisions of a local zoning law for a defined green benefit.	7
(3)	Place-based amenities such as plazas, squares, and attached greens located around civic, commercial, and mixed-use property are accessible by sidewalks, on-street parking, or provide for bike racks for the purpose of promoting higher density living.	7

405.5 Wetlands. Constructed wetlands or other natural innovative wastewater or stormwater treatment technologies are used.	8

405.6 Multi-modal transportation. Multi-modal transportation access is provided in accordance with one or more of the following:		
(1)	A site is selected with a boundary within one-half mile (805 m) of pedestrian access to a mass transit system or within five miles of a mass transit station with available parking.	5
(2)	A site is selected where all lots within the site are located within one-half mile (805 m) of pedestrian access to a mass transit system.	7
(3)	Walkways, bikeways, street crossings, and entrances designed to promote pedestrian activity are provided. New buildings are connected to existing sidewalks and areas of development.	5
(4)	Bicycle parking and racks are indicated on the site plan and constructed for mixed-use, multi-family buildings, and/or common areas.	4
(5)	Bike sharing programs participate with the developer and facilities for bike sharing are planned for and constructed.	5
(6)	Car sharing programs participate with the developer and facilities for car sharing are planned for and constructed.	5

405.7 Density. The average density on a net developable area basis is:		
(1)	7 to less than 14 dwelling units per acre (per 4,047 m^2)	5
(2)	14 to less than 21 dwelling units per acre (per 4,047 m^2)	7
(3)	21 or greater dwelling units per acre (per 4,047 m^2)	10

405.8 Mixed-use development. (1) Mixed-use development is incorporated, or (2) for single-use sites 20 acres or less in size with boundaries adjacent to a site with a minimum of two uses containing retail, services, and employment where a pedestrian network of streets, sidewalks, pathways, or plazas exists that connects a majority of lots within the site with the adjacent non-residential multi-use site.	9

GREEN BUILDING PRACTICES	POINTS
405.9 Open space. A portion of the gross area of the community is set aside as open space. **(Points awarded for every 10 percent of the community set aside as open space)**	**5**
405.10 Community garden(s). A portion of the site is established as a community garden(s) for the residents of the site to provide local food production for residents or area consumers.	**3**

CHAPTER 5

LOT DESIGN, PREPARATION, AND DEVELOPMENT

GREEN BUILDING PRACTICES	POINTS

500
LOT DESIGN, PREPARATION, AND DEVELOPMENT

500.0 Intent. This section applies to lot development for the eventual construction of residential buildings, multi-unit buildings, or additions thereto that contain dwelling units.

501 LOT
SELECTION

501.1 Lot. The lot is selected to minimize environmental impact by one or more of the following:

(1)	A lot is selected within a site certified to this Standard or equivalent.	6
(2)	An infill lot is selected.	8
(3)	An infill lot is selected that is a greyfield.	7
(4)	An EPA-recognized brownfield lot is selected.	9
(5)	A lot with an average slope calculation of less than 15% is selected.	9

501.2 Multi-modal transportation. A range of multi-modal transportation choices are promoted by one or more of the following:

(1)	A lot is selected within one-half mile (805 m) of pedestrian access to a mass transit system or within five miles (8,046 m) of a mass transit station with provisions for parking.	4
(2)	Walkways, street crossings, and entrances designed to promote pedestrian activity are provided. New buildings are connected to existing sidewalks and areas of development.	5
(3)	A lot is selected within one-half mile (805 m) of six or more community resources (e.g., recreational facilities (such as pools, tennis courts, basketball courts), parks, grocery store, post office, place of worship, community center, daycare center, bank, school, restaurant, medical/dental office, Laundromat/dry cleaner)].	4
(4)	Bicycle use is promoted by building on a lot located within a community that has rights-of-way specifically dedicated to bicycle use in the form of paved paths or bicycle lanes, or on an infill lot located within 1/2 mile of a bicycle lane designated by the jurisdiction.	5

502
PROJECT TEAM, MISSION STATEMENT, AND GOALS

502.1 Project team, mission statement, and goals. A knowledgeable team is established and team member roles are identified with respect to green lot design, preparation, and development. The project's green goals and objectives are written into a mission statement.	4

GREEN BUILDING PRACTICES	POINTS

503 LOT DESIGN

503.0 Intent. The lot is designed to avoid detrimental environmental impacts first, to minimize any unavoidable impacts, and to mitigate for those impacts that do occur. The project is designed to minimize environmental impacts and to protect, restore, and enhance the natural features and environmental quality of the lot.

(Points awarded only if the intent of the design is implemented.)

503.1 Natural resources. Natural resources are conserved by one or more of the following:	
(1)	A natural resources inventory is completed under the direction of a qualified professional.	5
(2)	A plan is implemented to conserve the elements identified by the resource inventory as high-priority resources.	6
(3)	Items listed for protection in the resource inventory plan are protected under the direction of a qualified professional.	4
(4)	Basic training in tree or other natural resource protection is provided for the on-site supervisor.	4
(5)	All tree pruning on-site is conducted by a Certified Arborist.	3
(6)	Ongoing maintenance of vegetation on the lot during construction is in accordance with TCIA A300 or locally accepted best practices.	4
(7)	Where a lot adjoins a landscaped common area, a protection plan from construction activities next to the common area is implemented.	5

503.2 Slope disturbance. Slope disturbance is minimized by one or more of the following:	
(1)	The use of terrain adaptive architecture including terracing, retaining walls, landscaping, or other restabilization techniques.	5
(2)	Hydrological/soil stability study is completed and used to guide the design of all buildings on the lot.	4
(3)	All or a percentage of driveways and parking are aligned with natural topography to reduce cut and fill.	
	(a) 10 percent to 25 percent	3
	(b) 25 percent to 75 percent	4
	(c) greater than 75 percent	6
(4)	Long-term erosion effects are reduced through the design and implementation of terracing, retaining walls, landscaping, or restabilization techniques.	5
(5)	Underground parking uses the natural slope for parking entrances.	5

GREEN BUILDING PRACTICES	POINTS
503.3 Soil disturbance and erosion. Soil disturbance and erosion are minimized by one or more of the following: (also see Section 504.3)	
(1) Construction activities are scheduled to minimize length of time that soils are exposed.	5
(2) At least 75% of total length of the utilities on the lot are designed to use one or more alternative means:	5
(a) tunneling instead of trenching	
(b) use of smaller (low ground pressure) equipment or geomats to spread the weight of construction equipment	
(c) shared utility trenches or easements	
(d) placement of utilities under paved surfaces instead of yards	
(3) Limits of clearing and grading are demarcated on the lot plan.	5
503.4 Stormwater management. Stormwater management includes one or more of the following low-impact development techniques: **(For lots in a development, the points for items (1), (2), and (3) may be awarded for the lot when there is a community stormwater management plan implemented and the builder does not violate that plan with respect to water leaving the lot.)**	
(1) Natural water and drainage features are preserved and used.	6
(2) Facilities that minimize concentrated flows and simulate flows found in natural hydrology by the use of vegetative swales, french drains, wetlands, drywells, rain gardens, or similar infiltration features.	7
(3) All or a percentage of impervious surfaces are minimized and permeable materials are used for driveways, parking areas, walkways, and patios.	
(a) less than 25 percent	2
(b) 25 percent to 75 percent	4
(c) greater than 75 percent	6
(4) A minimum of 50 percent of the roof is vegetated (green roof) using technology capable of withstanding the climate conditions of the jurisdiction and the microclimate conditions of the building lot. Invasive plant species are not permitted.	5
(5) Stormwater management practices manage rainfall on the lot and prevent the off-lot discharge from all storms up to and including the volume of the 95th percentile storm event.	6
(6) A hydrologic analysis is conducted that results in the design of a stormwater management system that maintains the pre-development (i.e., stable, natural) runoff hydrology of the lot throughout the development or redevelopment process. Post-construction runoff rate, volume, and duration cannot exceed predevelopment rates.	7

GREEN BUILDING PRACTICES	POINTS
503.5 Landscape plan. A landscape plan for the lot is developed to limit water and energy use while preserving or enhancing the natural environment. **(Where "front" only or "rear" only plan is implemented, only half of the points (rounding down to a whole number) are awarded for Items (1)-(6)**	
(1) Where a lot is less than 50 percent turf, a plan is formulated to restore or enhance natural vegetation that is cleared during construction. Landscaping is phased to coincide with achievement of final grades to ensure denuded areas are quickly vegetated.	6
(2) Turf grass species, other vegetation, and trees that are native or regionally appropriate for local growing conditions are selected and specified on the lot plan.	4
(3) The percentage of turf areas that is designed to be mowed is limited and shown on the lot plan. The percentage is based on the landscaped area of the lot not including the home footprint, hardscape, and any undisturbed natural areas.	
(a) 0 percent or EPA WaterSense Water Budget Tool is used to determine the maximum percentage of turf areas	5
(b) greater than 0 percent to less than 20 percent	4
(c) 20 percent to less than 40 percent	3
(d) 40 percent to 60 percent	2
(4) Plants with similar watering needs are grouped (hydrozoning) and shown on the lot plan.	5
(5) Summer shading by planting installed to shade a minimum of 30 percent of building walls. To conform to summer shading, the effective shade coverage (five years after planting) is the arithmetic mean of the shade coverage calculated at 10 am for eastward facing walls, noon for southward facing walls, and 3 pm for westward facing walls on the summer solstice.	5
(6) Vegetative wind breaks or channels are designed to protect the lot and immediate surrounding lots as appropriate for local conditions.	4
(7) Site or community generated tree trimmings or stump grinding of regionally appropriate trees are used on the lot to provide protective mulch during construction or for landscaping.	3
(8) An integrated pest management plan is developed to minimize chemical use in pesticides and fertilizers.	4

GREEN BUILDING PRACTICES	POINTS
503.6 Wildlife habitat. Measures are planned to support wildlife habitat and include at least two of the following:	
(1) Plants and gardens that encourage wildlife, such as bird and butterfly gardens.	3
(2) Inclusion of a certified "backyard wildlife" program.	3
(3) The lot is adjacent to a wildlife corridor, fish and game park, or preserved areas and is designed with regard for this relationship.	3
(4) Outdoor lighting techniques are utilized with regard for wildlife.	3

GREEN BUILDING PRACTICES	POINTS
503.7 Environmentally sensitive areas. The lot is in accordance with one or both of the following:	
(1) The lot does not contain any environmentally sensitive areas that are disturbed by the construction.	4
(2) Compromised environmentally sensitive areas are mitigated or restored.	4

       ICC 700-2012 NATIONAL GREEN BUILDING STANDARD™

GREEN BUILDING PRACTICES	POINTS

504 LOT CONSTRUCTION

504.0 Intent. Environmental impact during construction is avoided to the extent possible; impacts that do occur are minimized and any significant impacts are mitigated.

504.1 On-site supervision and coordination. On-site supervision and coordination is provided during on-the-lot clearing, grading, trenching, paving, and installation of utilities to ensure that specified green development practices are implemented. (also see Section 503.3)	4

504.2 Trees and vegetation. Designated trees and vegetation are preserved by one or more of the following:

(1)	Fencing or equivalent is installed to protect trees and other vegetation.	3
(2)	Trenching, significant changes in grade, and compaction of soil and critical root zones in all "tree save" areas as shown on the lot plan are avoided.	5
(3)	Damage to designated existing trees and vegetation is mitigated during construction through pruning, root pruning, fertilizing, and watering.	4

504.3 Soil disturbance and erosion implementation. On-site soil disturbance and erosion are minimized by one or more of the following in accordance with the SWPPP or applicable plan: (also see Section 503.3)

(1)	Sediment and erosion controls are installed on the lot and maintained in accordance with the stormwater pollution prevention plan, where required.	5
(2)	Limits of clearing and grading are staked out on the lot.	5
(3)	"No disturbance" zones are created using fencing or flagging to protect vegetation and sensitive areas on the lot from construction activity.	5
(4)	Topsoil from either the lot or the site development is stockpiled and stabilized for later use and used to establish landscape plantings on the lot.	5
(5)	Soil compaction from construction equipment is reduced by distributing the weight of the equipment over a larger area (laying lightweight geogrids, mulch, chipped wood, plywood, OSB, metal plates, or other materials capable of weight distribution in the pathway of the equipment).	4
(6)	Disturbed areas on the lot that are complete or to be left unworked for 21 days or more are stabilized within 14 days using methods as recommended by the EPA or in the approved SWPPP, where required.	3
(7)	Soil is improved with organic amendments or mulch.	3
(8)	Utilities on the lot are installed using one or more alternative means (e.g., tunneling instead of trenching, use of smaller equipment, use of low ground pressure equipment, use of geomats, shared utility trenches or easements).	5
(9)	Inspection reports of stormwater best management practices are available.	3

GREEN BUILDING PRACTICES	POINTS

505
INNOVATIVE PRACTICES

505.0 Intent. Innovative lot design, preparation, and development practices are used to enhance environmental performance. Waivers or variances from local development regulations are obtained and innovative zoning is used to implement such practices.

		POINTS
505.1 Driveways and parking areas. Driveways and parking areas are minimized by one or more of the following:		
(1)	Off-street parking areas are shared or driveways are shared. Waivers or variances from local development regulations are obtained to implement such practices, if required.	5
(2)	In a multi-unit project, parking capacity does not exceed the local minimum requirements.	5
(3)	Structured parking is utilized to reduce the footprint of surface parking areas.	
	(a) 25 percent to less than 50 percent	4
	(b) 50 percent to 75 percent	5
	(c) greater than 75 percent	6

		POINTS
505.2 Heat island mitigation. Heat island effect is mitigated by one or both of the following.		
(1)	Hardscape: Not less than 50 percent of the surface area of the hardscape on the lot meets one or a combination of the following methods.	5
	(a) Shading of hardscaping: Shade is provided from existing or new vegetation (within five years) or from trellises. Shade of hardscaping is to be measured on the summer solstice at noon.	
	(b) Light-colored hardscaping: Horizontal hardscaping materials are installed with a solar reflectance index (SRI) of 29 or greater. The SRI is calculated in accordance with ASTM E1980. A default SRI value of 35 for new concrete without added color pigment is permitted to be used instead of measurements.	
	(c) Permeable hardscaping: Permeable hardscaping materials are installed.	
(2)	Roofs: Not less than 75 percent of the exposed surface of the roof is in accordance with one or a combination of the following methods.	5
	(a) Minimum initial SRI of 78 for a low-sloped roof (a slope less than or equal to 2:12) and a minimum initial SRI of 29 for a steep-sloped roof (a slope of more than 2:12). The SRI is calculated in accordance with ASTM E1980. Roof products are certified and labeled.	
	(b) Roof is vegetated using technology capable of withstanding the climate conditions of the jurisdiction and the microclimate conditions of the building lot. Invasive plant species are not permitted.	

GREEN BUILDING PRACTICES	POINTS
505.3 Density. The average density on the lot on a net developable area basis is:	
(1) 7 to less than 14 dwelling units per acre (per 4,047 m^2)	5
(2) 14 to less than 21 dwelling units per acre (per 4,047 m^2)	8
(3) 21 or greater dwelling units per acre (per 4,047 m^2)	11

505.4 Mixed-use development. The lot contains a mixed-use building.	8

505.5 Community garden(s). A portion of the lot is established as a community garden(s), available to residents of the lot, to provide for local food production to residents or area consumers.	3

THIS PAGE INTENTIONALLY LEFT BLANK

CHAPTER 6

RESOURCE EFFICIENCY

GREEN BUILDING PRACTICES	POINTS

601
QUALITY OF CONSTRUCTION MATERIALS AND WASTE

601.0 Intent. Design and construction practices that minimize the environmental impact of the building materials are incorporated, environmentally efficient building systems and materials are incorporated, and waste generated during construction is reduced.

601.1 Conditioned floor area. Finished floor area of a dwelling unit is limited. Finished floor area is calculated in accordance with NAHBRC Z765. Only the finished floor area for stories above grade plane is included in the calculation.

(1)	less than or equal to 1,000 square feet (93 m^2)	**15**
(2)	less than or equal to 1,500 square feet (139 m^2)	**12**
(3)	less than or equal to 2,000 square feet (186 m^2)	**9**
(4)	less than or equal to 2,500 square feet (232 m^2)	**6**
(5)	greater than 4,000 square feet (372 m^2)	**Mandatory**
	(For every 100 square feet (9.29 m^2) over 4,000 square feet (372 m^2), one point is to be added to rating level points shown in Table 303, Category 7 for each rating level.)	
**Multi-Unit Building Note**: For a multi-unit building, a weighted average of the individual unit sizes is used for this practice.		

601.2 Material usage. Structural systems are designed or construction techniques are implemented that reduce and optimize material usage.	**9 Max**
(1) Minimum structural member or element sizes necessary for strength and stiffness in accordance with advanced framing techniques or structural design standards are selected.	3
(2) Higher-grade or higher-strength of the same materials than commonly specified for structural elements and components in the building are used and element or component sizes are reduced accordingly.	3
(3) Performance-based structural design is used to optimize lateral force-resisting systems.	3

601.3 Building dimensions and layouts. Building dimensions and layouts are designed to reduce material cuts and waste. This practice is used for a minimum of 80 percent of the following areas:	
(1) floor area	3
(2) wall area	3
(3) roof area	3
(4) cladding or siding area	3
(5) penetrations or trim area	1

GREEN BUILDING PRACTICES	POINTS
601.4 Framing and structural plans. Detailed framing or structural plans, material quantity lists and on-site cut lists for framing, structural materials, and sheathing materials are provided.	4
601.5 Prefabricated components. Precut or preassembled components, or panelized or precast assemblies are utilized for a minimum of 90 percent for the following system or building:	13 Max
(1) floor system	4
(2) wall system	4
(3) roof system	4
(4) modular construction for the entire building located above grade	13
(5) manufactured home construction for the entire building located above grade	13
601.6 Stacked stories. Stories above grade are stacked, such as in 1½-story, 2-story, or greater structures. The area of the upper story is a minimum of 50 percent of the area of the story below based on areas with a minimum ceiling height of 7 feet (2,134 mm).	8 Max
(1) first stacked story	4
(2) for each additional stacked story	2
601.7 Site-applied finishing materials. Building materials or assemblies listed below that do not require additional site-applied material for finishing are incorporated in the building.	12 Max
(a) pigmented, stamped, decorative, or final finish concrete or masonry	
(b) interior trim not requiring paint or stain	
(c) exterior trim not requiring paint or stain	
(d) window, skylight, and door assemblies not requiring paint or stain on one of the following surfaces: 　　i. exterior surfaces 　　ii. interior surfaces	
(e) interior wall coverings or systems not requiring paint or stain or other type of finishing application	
(f) exterior wall coverings or systems not requiring paint or stain or other type of finishing application	
(g) pre-finished hardwood flooring	
(1) 90 percent or more of the installed building materials or assemblies listed above:	5
(Points awarded for each type of material or assembly.)	
(2) 50 percent to less than 90 percent of the installed building material or assembly listed above:	2
(Points awarded for each type of material or assembly.)	
(3) 35 percent to less than 50 percent of the installed building material or assembly listed above:	1
(Points awarded for each type of material or assembly.)	

ICC 700-2012 NATIONAL GREEN BUILDING STANDARD™

GREEN BUILDING PRACTICES	POINTS
601.8 Foundations. A foundation system that minimizes soil disturbance, excavation quantities, and material usage, such as frost-protected shallow foundations, isolated pier and pad foundations, deep foundations, post foundations, or helical piles is selected, designed, and constructed. The foundation is used on 50 percent or more of the building footprint.	3
601.9 Above-grade wall systems. One or more of the following above-grade wall systems that provide sufficient structural and thermal characteristics are used for a minimum of 75 percent of the gross exterior wall area of the building:	4
(1) adobe	
(2) concrete and/or masonry	
(3) logs	
(4) rammed earth	

602
ENHANCED DURABILITY AND REDUCED MAINTENANCE

602.0 Intent. Design and construction practices are implemented that enhance the durability of materials and reduce in-service maintenance.	

602.1 Moisture Management – Building Envelope	

602.1.1 Capillary breaks	

602.1.1.1 A capillary break and vapor retarder are installed at concrete slabs in accordance with ICC IRC Sections R506.2.2 and R506.2.3 or ICC IBC Sections 1910 and 1805.4.1.	**Mandatory**
602.1.1.2 A capillary break between the footing and the foundation wall is provided to prevent moisture migration into foundation wall.	3
602.1.2 Foundation waterproofing. Enhanced foundation waterproofing is installed using one or both of the following:	4
(1) rubberized coating, or	
(2) drainage mat	

602.1.3 Foundation drainage	

602.1.3.1 Where required by the ICC IRC or IBC for habitable and usable spaces below grade, exterior drain tile is installed.	**Mandatory**
602.1.3.2 Interior and exterior foundation perimeter drains are installed and sloped to discharge to daylight, dry well, or sump pit.	4

GREEN BUILDING PRACTICES	POINTS

602.1.4 Crawlspaces

602.1.4.1 Vapor retarder in unconditioned vented crawlspace is in accordance with the following, as applicable. Joints of vapor retarder overlap a minimum of 6 inches (152 mm) and are taped.

(1)	Floors. Minimum 6 mil vapor retarder installed on the crawlspace floor and extended at least 6 inches up the wall and is attached and sealed to the wall.	6
(2)	Walls. Dampproof walls are provided below finished grade.	Mandatory

602.1.4.2 Crawlspace that is built as a conditioned area is sealed to prevent outside air infiltration and provided with conditioned air at a rate not less than 0.02 cfm (.009 L/s) per square foot of horizontal area and one of the following is implemented:

(1)	a concrete slab over 6 mil polyethylene or polystyrene sheeting lapped a minimum of 6 inches (152 mm) and taped or sealed at the seams.	8
(2)	6 mil polyethylene sheeting, lapped a minimum of 6 inches (152 mm), and taped at the seams.	Mandatory

602.1.5 Termite barrier. Continuous physical foundation termite barrier used with low toxicity treatment or with no chemical treatment is installed in geographical areas that have subterranean termite infestation potential determined in accordance with Figure 6(3).	4

602.1.6 Termite-resistant materials. In areas of termite infestation probability as defined by Figure 6(3), termite-resistant materials are used as follows:

(1)	In areas of slight to moderate termite infestation probability: for the foundation, all structural walls, floors, concealed roof spaces not accessible for inspection, exterior decks, and exterior claddings within the first 2 feet (610 mm) above the top of the foundation.	2
(2)	In areas of moderate to heavy termite infestation probability: for the foundation, all structural walls, floors, concealed roof spaces not accessible for inspection, exterior decks, and exterior claddings within the first 3 feet (914 mm) above the top of the foundation.	4
(3)	In areas of very heavy termite infestation probability: for the foundation, all structural walls, floors, concealed roof spaces not accessible for inspection, exterior decks, and exterior claddings.	6

602.1.7 Moisture control measures

602.1.7.1 Moisture control measures are in accordance with the following:

(1)	Building materials with visible mold are not installed or are cleaned or encapsulated prior to concealment and closing.	2
(2)	Insulation in cavities is dry in accordance with manufacturer's instructions when enclosed (e.g., with drywall).	Mandatory 2
(3)	The moisture content of lumber is sampled to ensure it does not exceed 19 percent prior to the surface and/or cavity enclosure.	4

602.1.7.2 Moisture content of subfloor, substrate, or concrete slabs is in accordance with the appropriate industry standard for the finish flooring to be applied.	2

GREEN BUILDING PRACTICES	POINTS
602.1.8 Water-resistive barrier. Where required by the ICC, IRC, or IBC, a water-resistive barrier and/or drainage plane system is installed behind exterior veneer and/or siding.	**Mandatory**

			POINTS
602.1.9 Flashing. Flashing is provided as follows to minimize water entry into wall and roof assemblies and to direct water to exterior surfaces or exterior water-resistive barriers for drainage. Flashing details are provided in the construction documents and are in accordance with the fenestration manufacturer's instructions, the flashing manufacturer's instructions, or as detailed by a registered design professional.			
(1)	Flashing is installed at all of the following locations, as applicable:		**Mandatory**
	(a)	around exterior fenestrations, skylights, and doors	
	(b)	at roof valleys	
	(c)	at all building-to-deck, -balcony, -porch, and -stair intersections	
	(d)	at roof-to-wall intersections, at roof-to-chimney intersections, at wall-to-chimney intersections, and at parapets.	
	(e)	at ends of and under masonry, wood, or metal copings and sills	
	(f)	above projecting wood trim	
	(g)	at built-in roof gutters, and	
	(h)	drip edge is installed at eaves and rake edges.	
(2)	All window head and jamb flashing is self-adhered flashing complying with AAMA 711-07.		2
(3)	Pan flashing is installed at sills of all exterior windows and doors.		3
(4)	Seamless, preformed kickout flashing, or prefabricated metal with soldered seams is provided at all roof-to-wall intersections. The type and thickness of the material used for roof flashing including but not limited kickout and step flashing is commensurate with the anticipated service life of the roofing material.		3
(5)	A rainscreen wall design as follows is used for exterior wall assemblies		4 Max
	(a)	a system designed with minimum 1/4-inch air space exterior to the water-resistive barrier, vented to the exterior at top and bottom of the wall, and integrated with flashing details. OR	4
	(b)	a cladding material or a water-resistive barrier with enhanced drainage, meeting 75 percent drainage efficiency determined in accordance with ASTM E2273.	2
(6)	Through-wall flashing is installed at transitions between wall cladding materials or wall construction types.		2
(7)	Flashing is installed at expansion joints in stucco walls		2

		POINTS
602.1.10 Exterior doors. Entries at exterior door assemblies, inclusive of side lights, are covered by one of the following methods to protect the building from the effects of precipitation and solar radiation. A projection factor of 0.375 minimum is provided. Eastern- and western-facing entries in Climate Zones 1, 2, and 3, as determined in accordance with Figure 6(1) or Appendix C, have a projection factor of 1.0 minimum, unless protected from direct solar radiation by other means (e.g., screen wall, vegetation).		**2 per exterior door** **6 Max**
	(a) installing a porch roof or awning	
	(b) extending the roof overhang	
	(c) recessing the exterior door	

GREEN BUILDING PRACTICES	POINTS
602.1.11 Tile backing materials. Tile backing materials installed under tiled surfaces in wet areas are in accordance with ASTM C1178, C1278, C1288, or C1325.	**Mandatory**

602.1.12 Roof overhangs. Roof overhangs, in accordance with Table 602.1.12, are provided over a minimum of 90 percent of exterior walls to protect the building envelope. — **4**

Table 602.1.12
Minimum Roof Overhang for One- & Two-Story Buildings

Inches of Rainfall [1]	Eave Overhang (Inches)	Rake Overhang (Inches)
≤40	12	12
>41 and ≤70	18	12
>70	24	12

(1) Annual mean total precipitation in inches is in accordance with Figure 6(2).

For SI: 12 inches = 304.8 mm

	POINTS
602.1.13 Ice barrier. In areas where there has been a history of ice forming along the eaves causing a backup of water, an ice barrier is installed in accordance with the ICC IRC or IBC at roof eaves of pitched roofs and extends a minimum of 24 inches (610 mm) inside the exterior wall line of the building.	**Mandatory**

602.1.14 Architectural features. Architectural features that increase the potential for water intrusion are avoided:	
(1) All horizontal ledgers are sloped away to provide gravity drainage as appropriate for the application.	**Mandatory 1**
(2) No roof configurations that create horizontal valleys in roof design.	2
(3) No recessed windows and architectural features that trap water on horizontal surfaces.	2

602.2 Roof surfaces. A minimum of 90 percent of roof surfaces, not used for roof penetrations and associated equipment, on-site renewable energy systems such as photovoltaics or solar thermal energy collectors, or rooftop decks, amenities and walkways, are constructed of one or both of the following:	3
(1) products that are in accordance with the ENERGY STAR® cool roof certification or equivalent	
(2) a vegetated roof system	

602.3 Roof water discharge. A gutter and downspout system or splash blocks and effective grading are provided to carry water a minimum of 5 feet (1524 mm) away from perimeter foundation walls.	4

602.4 Finished grade.	

602.4.1 Finished grade at all sides of a building is sloped to provide a minimum of 6 inches (150 mm) of fall within 10 feet (3048 mm) of the edge of the building. Where lot lines, walls, slopes, or other physical barriers prohibit 6 inches (152 mm) of fall within 10 feet (3048 mm), the final grade is sloped away from the edge of the building at a minimum slope of 2 percent.	**Mandatory**

GREEN BUILDING PRACTICES	POINTS
602.4.2 The final grade is sloped away from the edge of the building at a minimum slope of 5 percent.	1
602.4.3 Water is directed to drains or swales to ensure drainage away from the structure.	1

603 REUSED OR SALVAGED MATERIALS

GREEN BUILDING PRACTICES	POINTS
603.0 Intent. Practices that reuse or modify existing structures, salvage materials for other uses, or use salvaged materials in the building's construction are implemented.	
603.1 Reuse of existing building. Major elements or components of existing buildings and structures are reused, modified, or deconstructed for later use. **(Points awarded for every 200 square feet (18.5 m^2) of floor area.)**	1 12 Max
603.2 Salvaged materials. Reclaimed and/or salvaged materials and components are used. The total material value and labor cost of salvaged materials is equal to or exceeds 1 percent of the total construction cost. **(Points awarded per 1% of salvaged materials used based on the total construction cost.)** **(Materials, elements, or components awarded points under Section 603.1 shall not be awarded points under Section 603.2.)**	1 9 Max
603.3 Scrap materials. Sorting and reuse of scrap building material is facilitated (e.g., a central storage area or dedicated bins are provided).	4

604 RECYCLED-CONTENT BUILDING MATERIALS

GREEN BUILDING PRACTICES	POINTS
604.1 Recycled content. Building materials with recycled content are used for two minor and/or two major components of the building.	per Table 604.1

Table 604.1
Recycled Content

Material Percentage Recycled Content	Points Per 2 Minor	Points Per 2 Major
25% to less than 50%	1	2
50% to less than 75%	2	4
more than 75%	3	6

GREEN BUILDING PRACTICES	POINTS

605
RECYCLED CONSTRUCTION WASTE

605.0 Intent. Waste generated during construction is recycled. All waste classified as hazardous is properly handled and disposed of. **(Points not awarded for hazardous waste removal.)**	

605.1 Construction waste management plan. A construction waste management plan is developed, posted at the jobsite, and implemented with a goal of recycling or salvaging a minimum of 50 percent (by weight) of construction waste.	6

605.2 On-site recycling. On-site recycling measures following applicable regulations and codes are implemented, such as the following:	7
(a) Materials are ground or otherwise safely applied on-site as soil amendment or fill. A minimum of 50 percent (by weight) of construction and land-clearing waste is diverted from landfill.	
(b) Alternative compliance methods approved by the Adopting Entity.	
(c) Compatible untreated biomass material (lumber, posts, beams, etc.) are set aside for combustion if a solid fuel-burning appliance per Section 901.2.1(2) will be available for on-site renewable energy.	

605.3 Recycled construction materials. Construction materials (e.g., wood, cardboard, metals, drywall, plastic, asphalt roofing shingles, or concrete) are recycled offsite.	6 Max
(1) a minimum of two types of materials are recycled	3
(2) for each additional recycled material type	1

606
RENEWABLE MATERIALS

606.0 Intent. Building materials derived from renewable resources are used.	

606.1 Biobased products. The following biobased products are used:	8 Max
(a) certified solid wood in accordance with Section 606.2	
(b) engineered wood	
(c) bamboo	
(d) cotton	
(e) cork	
(f) straw	
(g) natural fiber products made from crops (soy-based, corn-based)	
(h) products with the minimum biobased contents of the USDA 7 CFR Part 2902	
(i) other biobased materials with a minimum of 50 percent biobased content (by weight or volume)	

GREEN BUILDING PRACTICES	POINTS
(1) Two types of biobased materials are used, each for more than 0.5 percent of the project's projected building material cost.	3
(2) Two types of biobased materials are used, each for more than 1 percent of the project's projected building material cost.	6
(3) For each additional biobased material used for more than 0.5 percent of the project's projected building material cost.	1 2 Max

606.2 Wood-based products. Wood or wood-based products are certified to the requirements of one of the following recognized product programs:	
(a) American Forest Foundation's *American Tree Farm System*® (ATFS)	
(b) Canadian Standards Association's *Sustainable Forest Management System Standards* (CSA Z809)	
(c) *Forest Stewardship Council* (FSC)	
(d) *Program for Endorsement of Forest Certification Systems* (PEFC)	
(e) *Sustainable Forestry Initiative*® *Program (SFI)*	
(f) other product programs mutually recognized by PEFC	
(1) A minimum of two certified wood-based products are used for minor elements of the building (e.g., all trim, cabinetry, or millwork).	3
(2) A minimum of two certified wood-based products are used in major elements of the building (e.g., walls, floors, roof).	4

606.3 Manufacturing energy. Materials manufactured using a minimum of 33 percent of the primary manufacturing process energy derived from (1) renewable sources, (2) combustible waste sources, or (3) renewable energy credits (RECs) are used for major components of the building. **(2 points awarded per material.)**	6 Max

607 RECYCLING AND WASTE REDUCTION

607.1 Recycling. Recycling by the occupant is facilitated by one or more of the following methods:	
(1) A built-in collection space in each kitchen and an aggregation/pick-up space in a garage, covered outdoor space, or other area for recycling containers is provided.	3
(2) Compost facility is provided on the site.	3

607.2 Food waste disposers. A minimum of one food waste disposer is installed at the primary kitchen sink.	1

GREEN BUILDING PRACTICES	POINTS

608 RESOURCE-EFFICIENT MATERIALS

608.1 Resource-efficient materials. Products containing fewer materials are used to achieve the same end-use requirements as conventional products, including but not limited to:	**9 Max 3 per each material**
(1) lighter, thinner brick with bed depth less than 3 inches and/or brick with coring of more that 25 percent	
(2) engineered wood or engineered steel products	
(3) roof or floor trusses	

609 REGIONAL MATERIALS

609.1 Regional materials. Regional materials are used for major elements or components of the building.	**10 Max 2 per each material type**

610 LIFE CYCLE ANALYSIS

610.1 Life cycle analysis. A life cycle analysis (LCA) tool is used to select environmentally preferable products or assemblies, or an LCA is conducted on the entire building. Points are awarded in accordance with Section 610.1.1 or 610.1.2. Only one method of analysis or tool may be utilized. The reference service life for the building is 60 years for any life cycle analysis tool. Results of the LCA are reported in the manual required in Section 1003.1(1) of this Standard in terms of the environmental impacts listed in this practice and it is stated if operating energy was included in the LCA.	**15 Max**

610.1.1 Whole-building life cycle analysis. A whole-building LCA is performed using a life cycle assessment and data compliant with ISO 14044 or other recognized standards.	**15**

610.1.2 Life cycle analysis for a product or assembly. An environmentally preferable product or assembly is selected for an application based upon the use of an LCA tool that incorporates data methods compliant with ISO 14044 or other recognized standards that compare the environmental impact of products or assemblies.	**10 Max**

GREEN BUILDING PRACTICES	POINTS

610.1.2.1 Product LCA. A product with improved environmental impact measures compared to another product(s) intended for the same use is selected. The environmental impact measures used in the assessment are selected from the following:	**per Table 610.1.2.1 10 Max**

    (a)   Fossil fuel consumption

    (b)   Global warming potential

    (c)   Acidification potential

    (d)   Eutrophication potential

    (e)   Ozone depletion potential

(Points are awarded for each product/system comparison where the selected product/system improved upon the environmental impact measures by an average of 15 percent.)

Table 610.1.2.1
Product LCA

4 Impact Measures	5 Impact Measures
POINTS	
2	3

610.1.2.2 Building assembly LCA. A building assembly with improved environmental impact measures compared to an alternative assembly of the same function is selected. The full life cycle, from resource extraction to demolition and disposal (including but not limited to on-site construction, maintenance and replacement, material and product embodied acquisition, and process and transportation energy), is assessed. The assessment includes all structural elements, insulation, and wall coverings of the assembly. The assessment does not include electrical and mechanical equipment and controls, plumbing products, fire detection and alarm systems, elevators, and conveying systems. The following types of building assemblies are eligible for points under this practice:	**per Table 610.1.2.2 10 Max**

    (a)   exterior walls

    (b)   roof/ceiling

    (c)   interior walls or ceilings

    (d)   intermediate floors

The environmental impact measures used in the assessment are selected from the following:

    (a)   Fossil fuel consumption

    (b)   Global warming potential

    (c)   Acidification potential

    (d)   Eutrophication potential

    (e)   Ozone depletion potential

(Points are awarded based on the number of types of building assemblies that improve upon environmental impact measures by an average of 15 percent.)

Table 610.1.2.2
Building Assembly LCA

Number of Types of Building Assemblies	4 Impact Measures	5 Impact Measures
	POINTS	
2 types	3	6
3 types	4	8
4 types	5	10

GREEN BUILDING PRACTICES	POINTS

611
INNOVATIVE PRACTICES

611.1 Manufacturer's environmental management system concepts. Product manufacturer's operations and business practices include environmental management system concepts, and the production facility is registered to ISO 14001 or equivalent. The aggregate value of building products from registered ISO 14001 or equivalent production facilities is 1 percent or more of the estimated total building materials cost. **(1 point awarded per percent.)**		**10 Max**

	611.2 Sustainable products. One or more of the following products are used for at least 30% of the floor or wall area of the entire dwelling unit, as applicable. Products are certified by a third-party agency accredited to ISO Guide 65.	**9 Max**
(1)	50% or more of carpet installed (by square feet) is certified to NSF 140.	3
(2)	50% or more of resilient flooring installed (by square feet) is certified to NSF 332.	3
(3)	50% or more of the insulation installed (by square feet) is certified to EcoLogo CCD-016.	3
(4)	50% or more of interior wall coverings installed (by square feet) is certified to NSF 342.	3
(5)	50% or more of the gypsum board installed (by square feet) is certified to ULE ISR 100.	3
(6)	50% or more of the door leafs installed (by number of door leafs) is certified to ULE ISR 102.	3
(7)	50% or more of the tile installed (by square feet) is certified to TCNA A138.1 Specifications for Sustainable Ceramic Tiles, Glass Tiles and Tile Installation Materials.	3

	611.3 Universal design elements. Dwelling incorporates one or more of the following universal design elements. Conventional industry construction tolerances are permitted.	**10 Max**
(1)	Any no-step entrance into the dwelling which (1) is accessible from a substantially level parking or drop-off area (no more than 2%) via an accessible path which has no individual change in elevation or other obstruction of more than 1-1/2 inches in height with the pitch not exceeding 1 in 12 and (2) provides a minimum 32-inch wide clearance into the dwelling.	3
(2)	Minimum 36-inch wide accessible route from the no-step entrance into at least one visiting room in the dwelling and into at least one full or half bathroom which has a minimum 32-inch clear door width and a 30-inch by 48-inch clear area inside the bathroom outside the door swing.	3
(3)	Minimum 36-inch wide accessible route from the no-step entrance into at least one bedroom which has a minimum 32-inch clear door width.	3
(4)	Blocking or equivalent installed in the accessible bathroom walls for future installation of grab bars at water closet and bathing fixture, if applicable.	1

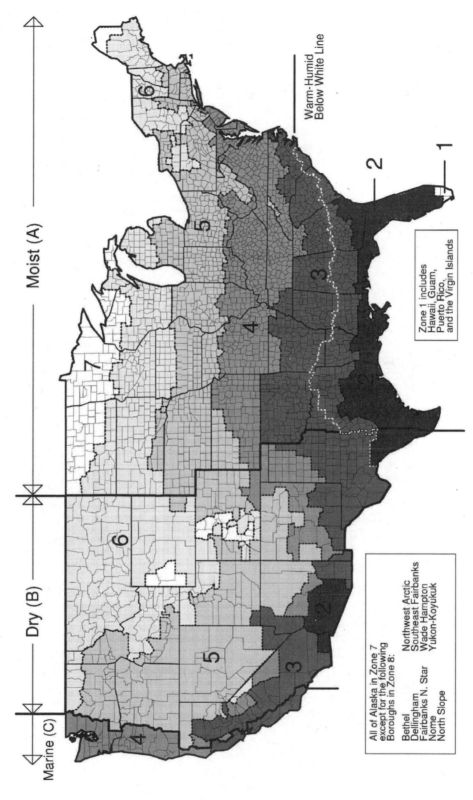

Moist (A)

Dry (B)

Marine (C)

Warm-Humid
Below White Line

2

1

Zone 1 includes
Hawaii, Guam,
Puerto Rico,
and the Virgin Islands

3

4

5

6

7

All of Alaska in Zone 7
except for the following
Boroughs in Zone 8:

Bethel   Northwest Arctic
Dellingham   Southeast Fairbanks
Fairbanks N. Star Wade Hampton
Nome   Yukon-Koyukuk
North Slope

FIGURE 6(1)
CLIMATE ZONES

Reprinted with permission from the 2009 International Residential Code, a copyrighted work of the International Code Council,
www.iccsafe.org.

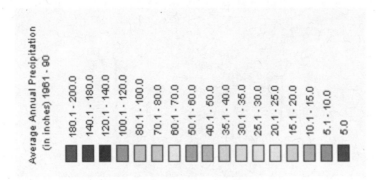

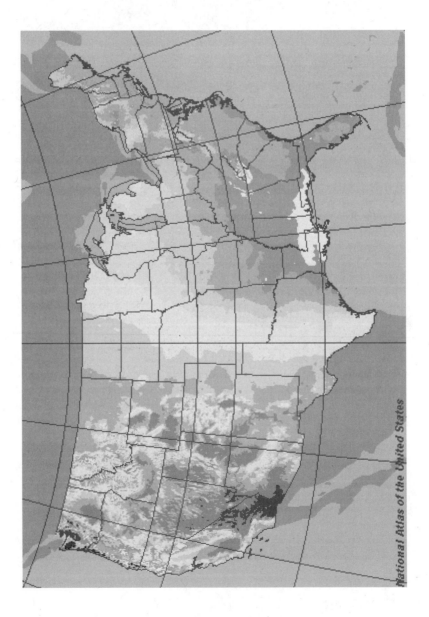

FIGURE 6(2)
AVERAGE ANNUAL PRECIPITATION (inches)

Source: www.nationalatlas.gov)

ICC 700-2012 NATIONAL GREEN BUILDING STANDARD™

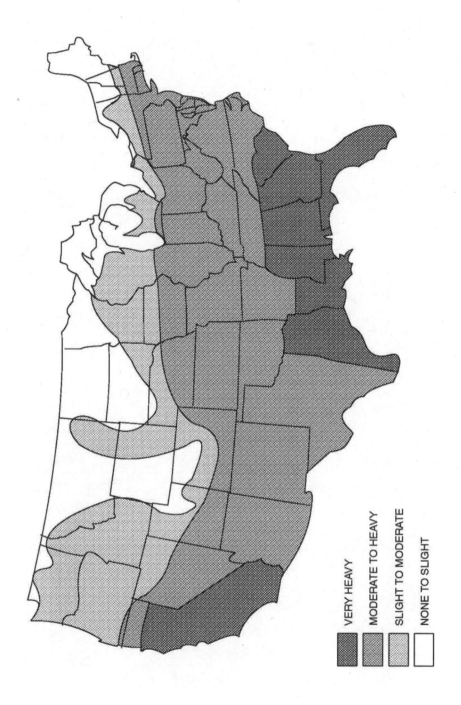

FIGURE 6(3)
TERMITE INFESTATION PROBABILITY MAP

Reprinted with permission from the 2009 International Residential Code, a copyrighted work of the International Code Council,
www.iccsafe.org.

THIS PAGE INTENTIONALLY LEFT BLANK

CHAPTER 7

ENERGY EFFICIENCY

GREEN BUILDING PRACTICES	POINTS

701
MINIMUM ENERGY EFFICIENCY REQUIREMENTS

701.1 Mandatory requirements. The building shall comply with either Section 702 (Performance Path) or Section 703 (Prescriptive Path). Items listed as "mandatory" in Section 701.4 apply to both the Performance and Prescriptive Paths.

701.1.1 Minimum Performance Path requirements. A building complying with Section 702 shall exceed the baseline minimum performance required by the ICC IECC by 15 percent and shall include a minimum of two practices from Section 704.

701.1.2 Minimum Prescriptive Path requirements. A building complying with Section 703 shall obtain a minimum of 30 points from Section 703 and shall include a minimum of two practices from Section 704.

701.1.3 Alternative bronze level compliance. As an alternative, any building that qualifies as an ENERGY STAR Version 3.0 Qualified Home or demonstrates compliance with the 2012 IECC or Chapter 11 of the 2012 IRC is deemed to meet all mandatory practices of Chapter 7 and achieves the bronze level for Chapter 7. The buildings achieving compliance under Section 701.1.3 are not eligible for achieving a rating level above bronze.

701.2 Emerald level points. The Performance Path shall be used to achieve the emerald level.

701.3 Adopting Entity review. A review by the Adopting Entity or designated third party shall be conducted to verify design and compliance with Chapter 7.

701.4 Mandatory practices.

701.4.1 HVAC systems.

701.4.1.1 HVAC system sizing. Space heating and cooling system is sized according to heating and cooling loads calculated using ACCA Manual J, or equivalent. Equipment is selected using ACCA Manual S or equivalent.	**Mandatory**
701.4.1.2 Radiant and hydronic space heating. Where installed as a primary heat source in the building, radiant or hydronic space heating system is designed using industry-approved guidelines and standards (e.g., ACCA Manual J, AHRI I=B=R, ACCA 5 QI-2010, or an accredited design professional's and manufacturer's recommendations).	**Mandatory**

GREEN BUILDING PRACTICES	POINTS

701.4.2 Duct systems.

701.4.2.1 Duct air sealing. Ducts are air sealed. All duct sealing materials are in conformance with UL 181A or UL 181B specifications and are installed in accordance with manufacturer's instructions.	Mandatory

701.4.2.2 Supply ducts. Building cavities are not used as supply ducts.	Mandatory

701.4.2.3 Duct system sizing. Duct system is sized and designed in accordance with ACCA Manual D or equivalent.	Mandatory

701.4.3 Insulation and air sealing.

701.4.3.1 Building Thermal Envelope. The building thermal envelope is durably sealed to limit infiltration. The sealing methods between dissimilar materials allow for differential expansion and contraction. The following are caulked, gasketed, weather-stripped or otherwise sealed with an air barrier material, suitable film, or solid material:	Mandatory
(a) All joints, seams and penetrations.	
(b) Site-built windows, doors, and skylights.	
(c) Openings between window and door assemblies and their respective jambs and framing.	
(d) Utility penetrations.	
(e) Dropped ceilings or chases adjacent to the thermal envelope.	
(f) Knee walls.	
(g) Walls and ceilings separating a garage from conditioned spaces.	
(h) Behind tubs and showers on exterior walls.	
(i) Common walls between dwelling units.	
(j) Attic access openings.	
(k) Rim joist junction.	
(l) Other sources of infiltration.	

701.4.3.2 Air sealing and insulation. Grade 3 insulation installation is not permitted. The compliance of the building envelope air tightness and insulation installation is demonstrated in accordance with Section 701.4.3.2(1) or 701.4.3.2(2).	Mandatory
(1) **Testing option.** Building envelope tightness and insulation installation is considered acceptable when air leakage is less than seven air changes per hour (ACH) when tested with a blower door at a pressure of 33.5 psf (50 Pa). Testing is conducted after rough-in and after installation of penetrations of the building envelope, including penetrations for utilities, plumbing, electrical, ventilation, and combustion appliances. Testing is conducted under the following conditions:	
(a) Exterior windows and doors, fireplace and stove doors are closed, but not sealed;	
(b) Dampers are closed, but not sealed, including exhaust, intake, makeup air, backdraft and flue dampers;	

GREEN BUILDING PRACTICES	POINTS

(c)	Interior doors are open;
(d)	Exterior openings for continuous ventilation systems and heat recovery ventilators are closed and sealed;
(e)	Heating and cooling systems are turned off;
(f)	HVAC duct terminations are not sealed; and
(g)	Supply and return registers are not sealed.

(2) **Visual inspection option.** Building envelope tightness and insulation installation are considered acceptable when the items listed in Table 701.4.3.2(2) applicable to the method of construction are field verified.

Table 701.4.3.2(2)
Air Barrier and Insulation Inspection Component Criteria

COMPONENT	CRITERIA
Air barrier and thermal barrier	• Exterior thermal envelope insulation for framed walls is installed in substantial contact and continuous alignment with building envelope air barrier. • Breaks or joints in the air barrier are filled or repaired. • Air-permeable insulation is not used as a sealing material. • Air-permeable insulation is installed with an air barrier.
Ceiling/attic	• Air barrier in dropped ceiling/soffit is substantially aligned with insulation and any gaps are sealed. • Attic access (except unvented attic), knee wall door, or drop-down stair is sealed.
Exterior walls	• Corners and headers are insulated. • Junction of foundation and sill plate is sealed.
Windows and doors	• Space between window/door jambs and framing is sealed.
Rim joists	• Rim joists are insulated and include an air barrier.
Floors (including above-garage and cantilevered floors)	• Insulation is installed to maintain permanent contact with underside of subfloor decking. • Air barrier is installed at any exposed edge of insulation.
Crawlspace walls	• Where installed, insulation is permanently attached to walls. • Exposed earth in unvented crawlspaces is covered with Class I vapor retarder with overlapping joints taped.
Shafts, penetrations	• Duct shafts, flue shafts, and utility penetrations opening to the exterior or an unconditioned space are sealed.
Narrow cavities	• Batts in narrow cavities are cut to fit, or narrow cavities are filled by sprayed/blown insulation.
Garage separation	• Air sealing is provided between the garage and conditioned spaces.
Recessed lighting	• Recessed light fixtures not installed in the conditioned space are air tight, IC rated, and sealed to drywall.
Plumbing and wiring	• Insulation is placed between the outside and pipes. Batt insulation is cut to fit around wiring and plumbing, or sprayed/blown insulation extends behind piping and wiring.
Shower/tub adjacent to exterior wall	• Showers and tubs adjacent to exterior walls have insulation and an air barrier separation from the exterior.
Electrical/phone box in exterior walls	• Air barrier extends behind boxes or air sealed-type boxes are installed.
Common wall	• Air barrier is installed in common walls between dwelling units.
HVAC register boots	• HVAC register boots that penetrate building envelope are sealed to subfloor or drywall.
Fireplace	• Fireplace walls include an air barrier.

GREEN BUILDING PRACTICES	POINTS
701.4.3.3 Fenestration air leakage. Windows, skylights and sliding glass doors have an air infiltration rate of no more than 0.3 cfm per square foot (1.5 L/s/m^2), and swinging doors no more than 0.5 cfm per square foot (2.6 L/s/m^2), when tested in accordance with NFRC 400 or AAMA/WDMA/CSA 101/I.S.2/A440 by an accredited, independent laboratory and listed and labeled. This practice does not apply to site-built windows, skylights, and doors.	**Mandatory**
701.4.3.4 Recessed lighting. Recessed luminaires installed in the building thermal envelope are sealed to limit air leakage between conditioned and unconditioned spaces. All recessed luminaires are IC-rated and labeled as meeting ASTM E283 when tested at 1.57 psf (75 Pa) pressure differential with no more than 2.0 cfm (0.944 L/s) of air movement from the conditioned space to the ceiling cavity. All recessed luminaires are sealed with a gasket or caulk between the housing and the interior of the wall or ceiling covering.	**Mandatory**
701.4.4 High-efficacy lighting. A minimum of 50 percent of the total hard-wired lighting fixtures, or the bulbs in those fixtures, qualify as high efficacy or equivalent.	**Mandatory**
701.4.5 Boiler supply piping. Boiler supply piping in unconditioned space is insulated.	**Mandatory**

702
PERFORMANCE PATH

702.1 Point allocation. Points from Section 702 (Performance Path) shall not be combined with points from Section 703 (Prescriptive Path).	**Mandatory**

702.2 Energy cost performance levels.

702.2.1 ICC IECC analysis. Energy efficiency features are implemented to achieve energy cost performance that meets the ICC IECC. A documented analysis using software in accordance with ICC IECC, Section 405, or ICC IECC Section 506.2 through 506.5, applied as defined in the ICC IECC, is required.	**Mandatory**

702.2.2 Energy cost performance analysis. Energy cost savings levels above the ICC IECC are determined through an analysis that includes improvements in building envelope, air infiltration, heating system efficiencies, cooling system efficiencies, duct sealing, water heating system efficiencies, lighting, and appliances.	
(1) 15 percent	**30**
(2) 30 percent	**60**
(3) 40 percent	**80**
(4) 50 percent	**100**

GREEN BUILDING PRACTICES	POINTS

703
PRESCRIPTIVE PATH

703.1 Building envelope

703.1.1 UA improvement. The total building thermal envelope UA is less than or equal to the total UA resulting from the U-factors provided in Table 703.1.1(a). Where insulation is used to achieve the UA improvement, the insulation installation is in accordance with Grade 1 requirements as graded by a third-party. Total UA is documented using a RESCheck or equivalent report to verify the baseline and the UA improvement.

Per Table 703.1.1(b)

Table 703.1.1(a)
Equivalent U-Factors[a]

Climate Zone	Fenestration U-Factor	Skylight U-Factor	Ceiling U-Factor	Frame Wall U-Factor	Mass Wall U-Factor[b]	Floor U-Factor	Basement Wall U-Factor	Crawlspace Wall U-Factor[c]
1	1.20	0.75	0.035	0.082	0.197	0.064	0.360	0.477
2	0.65	0.75	0.035	0.082	0.165	0.064	0.360	0.477
3	0.50	0.65	0.035	0.082	0.141	0.047	0.910	0.136
4 except Marine	0.35	0.60	0.030	0.082	0.141	0.047	0.059	0.065
5 and Marine 4	0.35	0.60	0.030	0.057	0.082	0.033	0.059	0.065
6	0.35	0.60	0.026	0.057	0.060	0.033	0.050	0.065
7 and 8	0.35	0.60	0.026	0.057	0.057	0.028	0.050	0.065

a. Non-fenestration U-factors shall be obtained from measurement, calculation, or an approved source.
b. Where more the half the insulation is on the interior, the mass wall U-factors is a maximum of 0.17 in Zone 1, 0.14 in Zone 2, 0.12 in Zone 3, 0.10 in Zone 4 except in Marine, and the same as the frame wall U-factor in Marine Zone 4 and Zones 5 through 8.
c. Basement wall U-factor of 0.360 in warm-humid locations.

Table 703.1.1(b)
Points for Improvement in Total Building Thermal Envelope UA

Minimum UA Improvement	Climate Zone							
	1	2	3	4	5	6	7	8
	POINTS							
0 to <5%	0	0	0	0	0	0	0	0
5% to <10%	0	2	3	4	7	5	3	4
10% to <15%	0	6	8	8	11	12	9	10
15% to <20%	0	10	12	13	16	14	11	12
≥20%	2	14	17	18	18	17	14	16

GREEN BUILDING PRACTICES	POINTS

703.1.2 Insulation installation. The insulation installation is graded by a third party and is in accordance with Sections 703.1.2.1, 703.1.2.2, and/or 703.1.2.3 as applicable. Grade 3 insulation installation is not permitted. Grade 2 installation is permitted only for bronze level buildings. **(Points not awarded in this section if already awarded under Section 703.1.1.)**	Per Table 703.1.2

Table 703.1.2
Insulation Installation Grades

Grade	POINTS
1	7
2	4

703.1.2.1 Grade 1 and Grade 2 insulation installations are in accordance with the following:

(1) Grading applies to field-installed insulation products.

(2) Grading applies to ceilings, walls, floors, band joists, rim joists, conditioned attics basements and crawlspaces, except as specifically noted.

(3) Inspection is conducted before insulation is covered.

(4) Air-permeable insulation is enclosed on all six sides and is in substantial contact with the sheathing material on one or more sides (interior or exterior) of the cavity. Air permeable insulation in ceilings is not required to be enclosed when the insulation is installed in substantial contact with the surfaces it is intended to insulate.

703.1.2.2 Grade 1 installation is in accordance with the following:

(1) Cavity insulation uniformly fills each cavity side-to-side and top-to-bottom, without substantial gaps or voids around obstructions (such as blocking or bridging).

(2) Cavity insulation compression or incomplete fill amounts to 2 percent or less, presuming the compressed or incomplete areas are a minimum of 70 percent of the intended fill thickness; occasional small gaps are acceptable.

(3) Exterior rigid insulation has substantial contact with the structural framing members or sheathing materials and is tightly fitted at joints.

(4) Cavity insulation is split, installed, and/or fitted tightly around wiring and other services.

(5) Exterior sheathing is not visible from the interior through gaps in the cavity insulation.

(6) Faced batt insulation is permitted to have side-stapled tabs, provided the tabs are stapled neatly with no buckling, and provided the batt is compressed only at the edges of each cavity, to the depth of the tab itself.

(7) Where properly installed, ICFs, SIPs, and other wall systems that provide integral insulation are deemed in compliance with the Grade 1 insulation installation requirements.

(8) Grade 1 insulation meets or exceeds all requirements for Grade 2 insulation.

703.1.2.3 Grade 2 installation is in accordance with the following:

(1) A maximum of 2 percent of the surface area of insulation is missing. Compression or incomplete fill amounts to 10 percent or less, presuming the compressed or incomplete areas are a minimum of 70 percent of the intended fill thickness.

GREEN BUILDING PRACTICES	POINTS

(2)	In unconditioned basements or unconditioned crawlspaces insulation is installed in substantial contact with the subfloor surfaces.
(a)	floor insulation over vented or ambient conditions is enclosed on six sides.
(b)	floor insulation over unconditioned basements is not required to be enclosed on six sides.
(3)	Ceiling insulation is not required to be enclosed when the insulation is installed in substantial contact with the drywall or plywood surfaces it is intended to insulate.
(4)	Eave baffles or equivalent construction is installed to prevent wind intrusion.
(5)	Installation with occasional installation defects is permitted: gaps around wiring, electrical outlets, plumbing and other intrusions; rounded edges or shoulders.

703.1.3 Mass walls. More than 75 percent of the above-grade exterior opaque wall area of the building is mass walls.

Per Table 703.1.3

Table 703.1.3
Exterior Mass Walls

Mass wall thickness	Climate Zone			
	1-4	5	6	7-8
	POINTS			
≥3 inch to <6 inch	5	4	3	0
>6 inch	3	2	2	0

703.1.4 A radiant barrier with an emittance of 0.05 or less is used in the attic. The product is tested in accordance with ASTM C1371 and installed in accordance with the manufacturer's instructions.

Per Table 703.1.4

Table 703.1.4
Radiant Barriers

Climate Zone	POINTS
1	2
2-3	3
4	1
5-8	0

703.1.5 Building envelope leakage. The maximum building envelope leakage rate is in accordance with Table 703.1.5 and whole building ventilation is provided in accordance with Section 902.2.1.

Per Table 703.1.5

Table 703.1.5
Building Envelope Leakage

Max Envelope Leakage Rate (ACH50)	Climate Zone							
	1	2	3	4	5	6	7	8
	POINTS							
5	2	3	3	4	6	7	8	9
4	3	4	5	7	10	12	13	14
3	3	5	6	9	13	15	17	19
2	4	6	8	11	15	18	20	23
1	4	5	8	12	17	19	22	24

GREEN BUILDING PRACTICES	POINTS

703.1.6 Fenestration

703.1.6.1 NFRC-certified (or equivalent) U-factor and SHGC of windows, exterior doors, skylights, and tubular daylighting devices (TDDs) on an area-weighted average basis are in accordance with Table 703.1.6.1. Area weighted averages are calculated separately for the categories of 1) windows and exterior doors and 2) skylights and tubular daylighting devices (TDDs). Decorative fenestration elements with a combined total maximum area of 15 square feet (1.39 m²) or 10 percent of the total glazing area, whichever is less, are not required to comply with this practice.	**Mandatory**

Table 703.1.6.1
Fenestration Specifications

Climate Zones	U-Factor	SHGC
	Windows and Exterior Doors (maximum certified ratings)	
1	0.65	0.30
2	0.65	0.30
3	0.40	0.30
4 to 8	0.35	Any
	Skylights and TDDs (maximum certified ratings)	
1 and 2	0.75	0.30
3	0.65	0.30
4 to 8	0.60	Any

703.1.6.2 The NFRC-certified (or equivalent) U-factor and SHGC of windows, exterior doors, skylights, and tubular daylighting devices (TDDs) are in accordance with Table 703.1.6.2(a), (b), or (c). Decorative fenestration elements with a combined total maximum area of 15 square feet (1.39 m²) or 10 percent of the total glazing area, whichever is less, are not required to comply with this practice.	**Per Table 703.1.6.2(a) or Table 703.1.6.2(b) or Table 703.1.6.2(c)**

Table 703.1.6.2(a)
Enhanced Fenestration Specifications

Climate Zones	U-Factor Windows & Exterior Doors	SHGC Windows & Exterior Doors	U-Factor Skylights & TDD's	SHGC Skylights & TDD's	POINTS
1	0.60	0.27	0.70	0.30	10
2	0.60	0.27	0.70	0.30	5
3	0.35	0.30	0.57	0.30	6
4	0.32	0.40	0.55	0.40	2
5	0.30	Any	0.55	Any	5
6	0.30	Any	0.55	Any	5
7	0.30	Any	0.55	Any	5
8	0.30	Any	0.55	Any	5

For Climate Zones 5-8 an equivalent energy performance is permitted based on either (1) windows with a U-factor = 0.31 and an SHGC ≥ 0.35, or, a U-factor = 0.32 and an SHGC ≥ 0.40 or (2) fenestration meeting the ENERGY STAR Equivalent Energy Performance requirements.

GREEN BUILDING PRACTICES	POINTS

Table 703.1.6.2(b)
Enhanced Fenestration Specifications

Climate Zone	U-Factor Windows & Exterior Doors	SHGC Windows & Exterior Doors	U-Factor Skylights & TDD's	SHGC Skylights & TDD's	POINTS
1	0.40	0.25	0.50	0.30	13
2	0.40	0.25	0.50	0.30	9
3	0.30	0.25	0.50	0.35	9
4	0.28	0.40	0.50	0.40	4
5	0.25	Any	0.50	Any	8
6	0.25	Any	0.50	Any	9
7	0.25	Any	0.50	Any	9
8	0.25	Any	0.50	Any	9

Table 703.1.6.2(c)
Enhanced Fenestration Specifications

Climate Zones	U-Factor Windows & Exterior Doors	SHGC Windows & Exterior Doors	U-Factor Skylights & TDD's	SHGC Skylights & TDD's	POINTS
4	0.25	0.40	0.40	0.40	5
5	0.22	Any	0.40	Any	9

703.2 HVAC equipment efficiency

703.2.1 Combination space heating and water heating system (combo system) is installed using either a coil from the water heater connected to an air handler to provide heat for the building or dwelling unit, or a space heating boiler using an indirect-fired water heater. Devices have a combined annual efficiency of 0.80.	4

703.2.2 Furnace and/or boiler efficiency is in accordance with one of the following:

(Where multiple systems are used, points awarded based on the system with the lowest efficiency.)

(1) Gas and propane heaters:	Per Table 703.2.2(1)

Table 703.2.2(1)
Gas and Propane Heaters

AFUE	Climate Zone							
	1	2	3	4	5	6	7	8
	POINTS							
≥90% AFUE	0	5	6	7	9	9	10	10
≥92% AFUE	0	5	8	9	11	11	12	12
≥94% AFUE	0	5	8	10	13	13	13	14
≥96% AFUE	1	6	10	11	14	14	15	16
≥98% AFUE	1	6	10	13	15	15	16	17

GREEN BUILDING PRACTICES	POINTS

(2) Oil furnace:

Per Table 703.2.2(2)

Table 703.2.2(2)
Oil Furnace

AFUE	Climate Zone					
	1	2	3	4	5	6-8
	POINTS					
≥85% AFUE	0	1	3	3	7	7
≥90% AFUE	0	2	5	8	11	14

(3) Gas boiler:

Per Table 703.2.2(3)

Table 703.2.2(3)
Gas Boiler

AFUE	Climate Zone					
	1	2	3	4	5	6-8
	POINTS					
≥85% AFUE	0	9	16	18	17	16
≥90% AFUE	1	10	17	19	18	17
≥94% AFUE	1	10	18	19	19	17
≥96% AFUE	1	10	18	20	19	18

(4) Oil boiler:

Per Table 703.2.2(4)

Table 703.2.2(4)
Oil Boiler

AFUE	Climate Zone					
	1	2	3	4	5	6-8
	POINTS					
≥85% AFUE	0	9	16	18	17	16
≥90% AFUE	1	10	17	19	18	17

703.2.3 Heat pump heating efficiency is in accordance with Table 703.2.3. Refrigerant charge is verified for compliance with manufacturer's instructions.

Per Table 703.2.3

(Where multiple systems are used, points awarded based on the system with the lowest efficiency.)

Table 703.2.3
Heat Pump Heating

Efficiency	Climate Zone					
	1	2	3	4	5	6-8[a]
	POINTS					
8.2 HSPF (11.5 EER)	0	1	2	4	5	5
9.0 HSPF (12.5 EER)	0	3	6	9	12	12
9.5 HSPF	0	4	7	12	16	16
10.0 HSPF	1	4	9	15	19	19

a. Equipment designed to operate in cold climates is recommended to minimize use of resistance heat when installing a heat pump in Zones 6-8.

GREEN BUILDING PRACTICES	POINTS

703.2.4 Cooling efficiency is in accordance with Table 703.2.4. Refrigerant charge is verified for compliance with manufacturer's instructions.

(Where multiple systems are used, points awarded based on the system with the lowest efficiency.)

Per Table 703.2.4

Table 703.2.4
Air Conditioner and Heat Pump Cooling

Efficiency	Climate Zone						
	1	2	3	4	5	6	7-8
	POINTS						
≥14 SEER (11.5 EER)	4	3	1	1	0	0	0
≥15 SEER (12.5 EER)	7	5	2	1	1	0	0
≥17 SEER (12.5 EER)	12	8	4	2	1	1	0
≥19 SEER (12.5 EER)	16	11	6	3	2	1	0
≥21 SEER	19	14	7	3	2	1	0

703.2.5 Water source cooling and heating efficiency is in accordance with Table 703.2.5.

Per Table 703.2.5

Table 703.2.5
Water Source Cooling And Heating

Efficiency	Climate Zone					
	1	2	3	4	5	6-8
	POINTS					
≥15 EER, ≥4.0 COP	14	18	22	30	37	37

703.2.6 Ground source heat pump is installed by a Certified Geothermal Service Contractor in accordance with Table 703.2.6.

(Where multiple systems are used, points awarded based on the system with the lowest efficiency.)

Per Table 703.2.6

Table 703.2.6
Ground source heat pump[a]

Efficiency	Climate Zone					
	1	2	3	4	5	6-8
	POINTS					
14.1 EER 3.3 COP	12	14	16	22	27	27
15 EER 3.5 COP	14	16	19	25	31	31
16.2 EER 3.6 COP	17	18	20	27	33	33
24 EER 4.3 COP	29	28	29	35	42	42
28 EER 4.8 COP	32	32	32	40	47	47

a. The ground loop is sized to account for the ground conductance and the expected minimum incoming water temperature to achieve rated performance.

703.2.7 ENERGY STAR, or equivalent, ceiling fans are installed.

(Points awarded per building.)

1

GREEN BUILDING PRACTICES	POINTS

703.2.8 Whole-building or whole-dwelling unit fan(s) with insulated louvers and a sealed enclosure is installed. **(Points awarded per building.)** **Table 703.2.8** **Whole dwelling unit fan**	**Per Table 703.2.8**

Table 703.2.8
Whole dwelling unit fan

Climate Zone		
1-3	4-6	7-8
POINTS		
5	3	0

703.2.9 In multi-unit buildings, an advanced electric and fossil fuel submetering system is installed to monitor electricity and fossil fuel consumption for each unit. The device provides consumption information on a monthly or near real-time basis. The information is available to the occupants at a minimum on a monthly basis.	1

703.3 Duct Systems

703.3.1 All space heating is provided by a system(s) that does not include air ducts.	**Per Table 703.3.1**

Table 703.3.1
Ductless heating system

Climate Zone					
1	2	3	4	5	6-8
POINTS					
0	4	7	7	6	2

703.3.2 All space cooling is provided by a system(s) that does not include air ducts.	**Per Table 703.3.2**

Table 703.3.2
Ductless cooling system

Climate Zone					
1	2	3	4	5	6-8
POINTS					
10	7	3	1	0	0

703.3.3 Ductwork is in accordance with all of the following:	**Per Table 703.3.3**
(1) Building cavities are not used as return ductwork.	
(2) Heating and cooling ducts and mechanical equipment are installed within the conditioned building space.	
(3) Ductwork is not installed in exterior walls.	

Table 703.3.3
Ducts

Climate Zone					
1	2	3	4	5	6-8
POINTS					
11	11	11	8	4	3

GREEN BUILDING PRACTICES	POINTS

703.3.4 Duct Leakage. The entire central HVAC duct system, including air handlers and register boots, is tested by a third party for total leakage at a pressure differential of 0.1 inches w.g. (25 Pa) and maximum air leakage is equal to or less than 6 percent of the system design flow rate. — *Per Table 703.3.4*

Table 703.3.4
Duct Leakage

Ductwork location	Climate Zone					
	1	2	3	4	5	6-8
	POINTS					
ductwork *entirely outside* the building's thermal envelope	8	9	8	6	3	2
ductwork *entirely inside* the building's thermal envelope	3	3	3	2	1	1
ductwork *inside and outside* the building's thermal envelope	5	6	5	4	2	2

703.4 Water heating system

703.4.1 Water heater Energy Factor (EF) is in accordance with the following:

(Where multiple systems are used, points awarded based on the system with the lowest efficiency.)

(1) Gas water heating — *Per Table 703.4.1(1)(a) or 703.4.1(1)(b)*

Table 703.4.1(1)(a)
Gas Water Heating

Energy Factor	Climate Zone							
	1	2	3	4	5	6	7	8
	POINTS							
0.67 to <0.80	4	4	3	2	3	2	1	1
≥0.80	7	7	5	4	5	4	2	2

Table 703.4.1(1)(b)
Gas Water Heating
(Storage with input rate greater than 75,000 Btu/h
or instantaneous input rate greater than 200,000 Btu/h)

Thermal Efficiency	Climate Zone							
	1	2	3	4	5	6	7	8
	POINTS							
≥0.86	7	7	5	4	5	4	2	2

(2) Electric water heating — *Per Table 703.4.1(2)*

Table 703.4.1(2)
Electric Water Heating

Energy Factor or Thermal Efficiency	Climate Zone							
	1	2	3	4	5	6	7	8
	POINTS							
≥0.95	2	2	2	1	1	1	1	1

GREEN BUILDING PRACTICES	POINTS

(3) Oil water heating

Per Table 703.4.1(3)

Table 703.4.1(3)
Oil Water Heating

Size (gallons)	Energy Factor	Climate Zone							
		1	2	3	4	5	6	7	8
		POINTS							
30 to <50	0.59	1	1	1	1	1	1	1	1
≥50	0.59	1	1	1	1	1	1	1	1

For SI: 1 gallon = 3.785 L

(4) Heat pump water heating

Per Table 703.4.1(4)

Table 703.4.1(4)
Heat Pump Water Heating

Energy Factor	Climate Zone							
	1	2	3	4	5	6	7	8
	POINTS							
1.5 to <2.0	14	11	11	11	11	4	4	4
2.0 to <2.2	19	16	16	15	15	6	6	6
2.2	20	17	17	17	16	6	6	6

703.4.2 Desuperheater is installed by a qualified installer or is pre-installed in the factory.

Per Table 703.4.2

Table 703.4.2
Desuperheater

Climate Zone		
1	2-5	6-8
POINTS		
17	8	4

703.4.3 Drain-water heat recovery system is installed in multi-family units.

(Points awarded per building.)

2

703.4.4 Indirect-fired water heater storage tanks heated from boiler systems are installed.

1

703.4.5 Solar water heater. SRCC (Solar Rating & Certification Corporation) OG 300 rated, or equivalent, solar domestic water heating system is installed. Solar Energy Factor (SEF) as defined by SRCC is in accordance with Table 703.4.5.

Per Table 703.4.5

Table 703.4.5
Solar Hot Water Systems

SEF	Climate Zone							
	1	2	3	4	5	6	7	8
	POINTS							
SEF 1.3	15	10	11	12	12	10	7	4
SEF 1.51	18	12	14	14	15	12	8	5
SEF 1.81	21	14	16	17	18	14	10	6
SEF 2.31	24	17	19	20	22	16	12	7
SEF 3.01	27	19	21	23	25	18	13	8

GREEN BUILDING PRACTICES	POINTS

703.5 Lighting and appliances

703.5.1 Hard-wired lighting. Hard-wired lighting is in accordance with one of the following:

(1) A minimum percent of the total hard-wired luminaires qualify as ENERGY STAR or equivalent.	Per Table 703.5.1

Table 703.5.1
Hard-wired Lighting

Minimum percent of fixtures	Climate Zone							
	1	2	3	4	5	6	7	8
	POINTS							
75%	5	4	3	3	3	2	2	1
95%	9	6	5	4	4	3	2	1

(2) A minimum of 80 percent of the exterior lighting wattage has a minimum efficiency of 40 lumens per watt or is solar-powered.	1

703.5.2 Recessed luminaires. The number of recessed luminaires that penetrates the thermal envelope is less than 1 per 400 square feet (37.16 m^2) of total conditioned floor area and they are in accordance with Section 701.4.3.4.	2

703.5.3 Appliances. ENERGY STAR or equivalent appliance(s) are installed:

(1) Refrigerator	Per Table 703.5.3(1)

Table 703.5.3(1)
Refrigerator

Climate Zone							
1	2	3	4	5	6	7	8
POINTS							
3	2	1	1	1	1	1	1

(2) Dishwasher	1
(3) Washing machine	4

703.5.4 Induction cooktop. Induction cooktop is installed.	1

703.6 Passive solar design

703.6.1 Sun-tempered design. Building orientation, sizing of glazing, and design of overhangs are in accordance with all of the following:	5
(1) The long side (or one side if of equal length) of the building faces within 20 degrees of true south.	
(2) Vertical glazing area is between 5 and 7 percent of the gross conditioned floor area on the south face [also see Section 703.6.1(8)].	
(3) Vertical glazing area is less than 2 percent of the gross conditioned floor area on the west face, and glazing is ENERGY STAR compliant or equivalent.	

GREEN BUILDING PRACTICES	POINTS

(4)		Vertical glazing area is less than 4 percent of the gross conditioned floor area on the east face, and glazing is ENERGY STAR compliant or equivalent.
(5)		Vertical glazing area is less than 8 percent of the gross conditioned floor area on the north face, and glazing is ENERGY STAR compliant or equivalent.
(6)		Skylights, where installed, are in accordance with the following:
	(a)	shades and insulated wells are used, and all glazing is ENERGY STAR compliant or equivalent
	(b)	horizontal skylights are less than 0.5 percent of finished ceiling area
	(c)	sloped skylights on slopes facing within 45 degrees of true south, east, or west are less than 1.5 percent of the finished ceiling area
(7)		Overhangs or adjustable canopies or awnings or trellises provide shading on south-facing glass for the appropriate climate zone in accordance with Table 703.6.1(7):

Table 703.6.1(7)
South-Facing Window Overhang Depth

Climate Zone		Vertical distance between bottom of overhang and top of window sill				
		≤7' 4"	≤6' 4"	≤5' 4"	≤4' 4"	≤3' 4"
Climate Zone	1 & 2 & 3	2' 8"	2' 8"	2' 4"	2' 0"	2' 0"
	4 & 5 & 6	2' 4"	2' 4"	2' 0"	2' 0"	1' 8"
	7 & 8	2' 0"	1' 8"	1' 8"	1' 4"	1' 0"

For SI: 1 inch = 25.4 mm

(8)		The south face windows have a SHGC of 0.40 or higher.
(9)		Return air or transfer grilles/ducts are in accordance with Section 704.3.

703.6.2 Window shading. Automated solar protection is installed to provide shading for windows.	1

703.6.3 Passive cooling design. Passive cooling design features are in accordance with three or more of the following:	
Points for three items:	3
Points for one additional item:	1

(1)		Exterior shading is provided on east and west windows using one or a combination of the following:
	(a)	vine-covered trellises with the vegetation separated a minimum of 1 foot (305 mm) from face of building
	(b)	moveable awnings or louvers
	(c)	covered porches
	(d)	attached or detached conditioned/unconditioned enclosed space that provides full shade of east and west windows (e.g., detached garage, shed, or building)

GREEN BUILDING PRACTICES	POINTS

(2)	Overhangs are installed to provide shading on south-facing glazing in accordance with Section 703.6.1(7).	
	(Points not awarded if points are taken under Section 703.6.1.)	
(3)	Windows and/or venting skylights are located to facilitate cross ventilation.	
(4)	Solar reflective roof or radiant barrier is installed in climate zones 1, 2, or 3 and roof material achieves a 3-year aged criteria of 0.50.	
(5)	Internal exposed thermal mass is a minimum of three inches (76 mm) in thickness. Thermal mass consists of concrete, brick, and/or tile fully adhered to a masonry base or other masonry material in accordance with one or a combination of the following:	
	(a)	A minimum of 1 square foot (0.09 m^2) of exposed thermal mass of floor per 3 square feet (2.8 m^2) of gross finished floor area.
	(b)	A minimum of 3 square feet (2.8 m^2) of exposed thermal mass in interior walls or elements per square foot (0.09 m^2) of gross finished floor area.
(6)	Roofing material is installed with a minimum 0.75 inch (19 mm) continuous air space offset from the roof deck from eave to ridge.	

703.6.4 Passive solar heating design. In addition to the sun-tempered design features in Section 703.6.1, all of the following are implemented:	4

(1)	Additional glazing, no greater than 12 percent, is permitted on the south wall. This additional glazing is in accordance with the requirements of Section 703.6.1.
(2)	Additional thermal mass for any room with south-facing glazing of more than 7 percent of the finished floor area is provided in accordance with the following:
	(a) Thermal mass is solid and a minimum of 3 inches (76 mm) in thickness. Where two thermal mass materials are layered together (e.g., ceramic tile on concrete base) to achieve the appropriate thickness, they are fully adhered to (touching) each other.
	(b) Thermal mass directly exposed to sunlight is provided in accordance with the following minimum ratios:
	(i) Above latitude 35 degrees: 5 square feet (0.465 m^2) of thermal mass for every 1 square foot (0.0929 m^2) of south-facing glazing.
	(ii) Latitude 30 degrees to 35 degrees: 5.5 square feet (0.51 m^2) of thermal mass for every 1 square foot (0.0929 m^2) of south-facing glazing.
	(iii) Latitude 25 degrees to 30 degrees: 6 square feet (0.557 m^2) of thermal mass for every 1 square foot (0.0929 m^2) of south-facing glazing.
	(c) Thermal mass not directly exposed to sunlight is permitted to be used to achieve thermal mass requirements of Section 703.6.4 (2) based on a ratio of 40 square feet (3.72 m^2) of thermal mass for every 1 square foot (0.0929 m^2) of south-facing glazing.
(3)	In addition to return air or transfer grilles/ducts required by Section 703.6.1(9), provisions for forced airflow to adjoining areas are implemented as needed.

GREEN BUILDING PRACTICES	POINTS

704
ADDITIONAL PRACTICES

704.1 Application of additional practice points. Points from Section 704 can be added to points earned in Section 702 (Performance Path), Section 703 (Prescriptive Path), or Section 701.1.3 (alternative bronze level compliance).

704.2 Lighting

704.2.1 Occupancy sensors. Occupancy sensors are installed on indoor lights, and photo or motion sensors are installed on outdoor lights to control lighting.

(1)	25 percent of lighting	1
(2)	50 percent of lighting	2

704.2.2 TDDs and skylights. Tubular daylighting device (TDD) or a skylight with sealed, insulated, low-E glass is installed in rooms without windows. **(Points awarded per building.)**	2

704.2.3 Lighting outlets. Occupancy sensors are installed for a minimum of 80 percent of hard-wired lighting outlets.	1

704.3 Return ducts and transfer grilles. Return ducts or transfer grilles are installed in every room with a door. Return ducts or transfer grilles are not required for bathrooms, kitchens, closets, pantries, and laundry rooms.	5

704.4 HVAC design and installation

704.4.1 HVAC contractor and service technician are certified by a nationally or regionally recognized program (e.g., North American Technician Excellence, Inc. (NATE), Air Conditioning Contractors of Americas Quality Assured Program (ACCA/QA), Building Performance Institute (BPI), Radiant Panel Association, or a manufacturer's training program).	1

704.4.2 Performance of the heating and/or cooling system is verified by the HVAC contractor in accordance with all of the following:	3
(1) Start-up procedure is performed in accordance with the manufacturer's instructions.	
(2) Refrigerant charge is verified by super-heat and/or sub-cooling method.	
(3) Burner is set to fire at input level listed on nameplate.	
(4) Air handler setting/fan speed is set in accordance with manufacturer's instructions.	
(5) Total airflow is within 10 percent of design flow.	
(6) Total external system static does not exceed equipment capability at rated airflow.	

GREEN BUILDING PRACTICES	POINTS
704.4.3 Manufacturer's label or printed specifications for sealed air handler (except furnaces) indicates the leakage is less than or equal to 2 percent of design airflow at a pressure of 1 inch of water (250 Pa). Air handlers are tested with inlets, outlets, and condensate drain ports sealed and filter box in place.	4

704.5 Installation and performance verification.

704.5.1 Third-party on-site inspection is conducted to verify compliance with all of the following, as applicable. Minimum of two inspections are performed: one inspection after insulation is installed and prior to covering, and another inspection upon completion of the building. Where multiple buildings or dwelling units of the same model are built by the same builder, a representative sample inspection of a minimum of 15 percent of the buildings or dwelling units is permitted.	5
(1) Ducts are installed in accordance with the ICC IRC or IMC and ducts are sealed.	
(2) Building envelope air sealing is installed.	
(3) Insulation is installed in accordance with Section 703.1.2.	
(4) Windows, skylights, and doors are flashed, caulked, and sealed in accordance with manufacturer's instructions and in accordance with Section 701.4.3.	

704.5.2 Testing. Testing above mandatory requirements is conducted to verify performance.

704.5.2.1 Building envelope leakage testing.	
(1) A blower door test and a visual inspection are performed as described in 701.4.3.2.	5
(2) Third-party verification is completed.	5

704.5.2.2 HVAC airflow testing. Balanced HVAC airflows are demonstrated by flow hood or other acceptable flow measurement tool by a third party. Test results are in accordance with both of the following:	8
(1) Measured flow at each supply and return register is within 25 percent of design flow.	
(2) Total airflow is within 10 percent of design flow.	

704.5.3 Insulating hot water pipes. Insulation with a minimum thermal resistance (R-value) of at least R-3 is applied to the following, as applicable:	1
(a) piping larger than 3/4-inch outside diameter	
(b) piping serving more than one dwelling unit	
(c) piping branches serving kitchen sinks	
(d) piping located outside the conditioned space	
(e) piping from the water heater to a distribution manifold	
(f) piping located under a floor slab	
(g) buried piping	
(h) piping in recirculation systems other than demand recirculation systems	

GREEN BUILDING PRACTICES	POINTS

(i) all other piping except the piping that meets the length requirements of Table 704.5.3

Table 704.5.3
Maximum Pipe Run Length

Nominal Pipe Diameter of largest pipe in run (inches)	Maximum pipe length (feet)[a]
3/8	30
1/2	20
3/4	10

a. Total length of all piping from the source of hot water (either a water heater or distribution manifold (or tee) on a trunk line or a recirculation loop) to a point of use.

705
INNOVATIVE PRACTICES

705.1 Energy consumption control. A whole-building or whole-dwelling unit device is installed that controls or monitors energy consumption.	**7 Max**
(1) programmable communicating thermostat	1
(2) energy-monitoring device	2
(3) energy management control system	4

705.2 Renewable energy service plan. Renewable energy service plan is provided as follows:	
(1) Builder selects a renewable energy service plan provided by the local electrical utility for interim (temporary) electric service. The builder's local administrative office has renewable energy service.	1
(2) The buyer of the building selects one of the following renewable energy service plans provided by the utility prior to occupancy of the building with a minimum two-year commitment.	
(a) less than half of the dwelling's projected electricity and gas use is provided by renewable energy	1
(b) half or more of the of the dwelling's projected electricity and gas use is provided by renewable energy	5

705.3 Smart Appliances and Systems. Smart appliances and systems are installed as follows:	
(1) Refrigerator	
(2) Freezer	
(3) Dishwasher	
(4) Clothes Dryer	
(5) Clothes Washer	
(6) Room Air Conditioner	
(7) HVAC Systems	
(8) Service Hot Water Heating Systems	
Three to five smart appliances installed	1
Six or more smart appliances installed	2

GREEN BUILDING PRACTICES	POINTS

705.4 Pumps.

705.4.1 Pool, spa, and water features equipped with filtration pumps as follows:	
(1) Two-speed pump(s) is installed.	1
(2) Electronically controlled variable-speed pump(s) is installed (efficiency of 90 percent or greater).	3

705.4.2 Sump pump(s) with electrically commutated motors (ECMs) or permanent split capacitor (PSC) motors is installed (efficiency of 90 percent or greater).	1

705.5 Additional renewable energy options. Renewable energy system(s) is installed on the property (e.g., solar photovoltaic panels, building integrated photovoltaic system, wind energy system, on-site micro-hydro power system, active solar space heating system, solar thermal hydronic heating system, photovoltaic hybrid heating system). **(Points awarded per 100 W of system rating per 2,000 square feet of total conditioned floor area of the building.)**	1

705.6 Parking garage efficiency. Structured parking garages are designed to require no mechanical ventilation for fresh air requirements.	2

THIS PAGE INTENTIONALLY LEFT BLANK

CHAPTER 8

WATER EFFICIENCY

GREEN BUILDING PRACTICES	POINTS

801
INDOOR AND OUTDOOR WATER USE

801.0 Intent. Measures that reduce indoor and outdoor water usage are implemented.

801.1 Indoor hot water usage. Indoor hot water supply system is in accordance with one of the practices listed in items (1) through (5). The maximum water volume from the source of hot water to the termination of the fixture supply is determined in accordance with Tables 801.1(1) or 801.1(2). The maximum pipe length from the source of hot water to the termination of the fixture supply is 50 feet.

(Where more than one water heater is used or where more than one type of hot water supply system, including multiple circulation loops, is used, points are awarded only for the system that qualifies for the minimum number of points.)
(Systems with circulation loops are eligible for points only if pumps are demand controlled. Circulation systems with timers or aquastats and constant-on circulation systems are not eligible to receive points.)
(Points awarded only if the pipes are insulated in accordance with Section 704.5.3.)

		POINTS
(1)	The maximum volume from the water heater to the termination of the fixture supply at furthest fixture is 128 ounces (1 gallon or 3.78 liters).	11
(2)	The maximum volume from the water heater to the termination of the fixture supply at furthest fixture is 64 ounces (0.5 gallon or 1.89 liters).	17
(3)	The maximum volume from the water heater to the termination of the fixture supply at furthest fixture is 32 ounces (0.25 gallon or 0.945 liters).	29
(4)	A demand controlled hot water priming pump is installed on the main supply pipe of the circulation loop and the maximum volume from this supply pipe to the furthest fixture is 24 ounces (0.19 gallons or 0.71 liters).	35
	(a) The volume in the circulation loop (supply) from the water heater or boiler to the branch for the furthest fixture is no more than 128 ounces (1 gallon or 3.78 liters).	4 Additional
(5)	A central hot water recirculation system is implemented in multi-unit buildings in which the hot water line distance from the recirculating loop to the engineered parallel piping system (i.e., manifold system) is less than 30 feet (9,144 mm) and the parallel piping to the fixture fittings contains a maximum of 64 ounces (1.89 liters) (115.50 cubic inches) (0.50 gallons).	9

GREEN BUILDING PRACTICES	POINTS
(6) Tankless water heater(s) with at least 0.5 gallon (1.89 liters) of storage are installed, or a tankless water heater that ramps up to at least 110F within 5 seconds is installed. The storage may be internal or external to the tankless water heater.	**4 Additional**

Table 801.1(1)

Maximum Pipe Length Conversion Table[a]

Nominal Pipe Size (inch)	Liquid Ounces per Foot of Length	Main, Branch, and Fixture Supply System Volume Category			Branch and Fixture Supply Volume from Circulation Loop
		128 ounces (1 gallons) [per 801.1(1)]	64 ounces (0.5 gallon) [per 801.1(2)]	32 ounces (0.25 gallon) [per 801.1(3)]	24 ounces (0.19 gallon) [per 801.1(4)]
		Maximum Pipe Length (feet)			
1/4[b]	0.33	50	50	50	50
5/16[b]	0.5	50	50	50	48
3/8[b]	0.75	50	50	43	32
1/2	1.5	50	43	21	16
5/8	2	50	32	16	12
3/4	3	43	21	11	8
7/8	4	32	16	8	6
1	5	26	13	6	5
1 1/4	8	16	8	4	3
1 1/2	11	12	6	3	2
2	18	7	4	2	1

a. Maximum pipe length figures apply when the entire pipe run is one nominal diameter only. Where multiple pipe diameters are used, the combined volume shall not exceed the volume limitation in Section 801.1.

b. The maximum flow rate through 1/4 inch nominal piping shall not exceed 0.5 gpm. The maximum flow rate through 5/16 inch nominal piping shall not exceed 1 gpm. The maximum flow rate through 3/8 inch nominal piping shall not exceed 1.5 gpm.

Table 801.1(2)

Common Hot Water Pipe Internal Volumes

OUNCES OF WATER PER FOOT OF PIPE

Size Nominal, Inch	Copper Type M	Copper Type L	Copper Type K	CPVC CTS SDR 11	CPVC SCH 40	CPVC SCH 80	PE-RT SDR 9	Composite ASTM F 1281	PEX CTS SDR 9
3/8	1.06	0.97	0.84	N/A	1.17	N/A	0.64	0.63	0.64
1/2	1.69	1.55	1.45	1.25	1.89	1.46	1.18	1.31	1.18
3/4	3.43	3.22	2.90	2.67	3.38	2.74	2.35	3.39	2.35
1	5.81	5.49	5.17	4.43	5.53	4.57	3.91	5.56	3.91
1 ¼	8.70	8.36	8.09	6.61	9.66	8.24	5.81	8.49	5.81
1 ½	12.18	11.83	11.45	9.22	13.2	11.38	8.09	13.88	8.09
2	21.08	20.58	20.04	15.79	21.88	19.11	13.86	21.48	13.86

GREEN BUILDING PRACTICES	POINTS
801.2 Water-conserving appliances. ENERGY STAR or equivalent water-conserving appliances are installed.	
(1) dishwasher	2
(2) washing machine, or	13
(3) washing machine with a water factor of 6.0 or less	24
Multi-Unit Building Note: Washing machines are installed in individual units or provided in common areas of multi-unit buildings.	

GREEN BUILDING PRACTICES	POINTS
801.3 Showerheads. Showerheads are in accordance with the following:	
(1) The total maximum combined flow rate of all showerheads controlled by a single valve at any point in time in a shower compartment is 1.6 to less than 2.5 gpm. Maximum of two valves are installed per shower compartment. The flow rate is tested at 80 psi (552 kPa) in accordance with ASME A112.18.1. Showerheads are served by an automatic compensating valve that complies with ASSE 1016 or ASME A112.18.1 and specifically designed to provide thermal shock and scald protection at the flow rate of the showerhead. **(Points awarded per shower compartment. In multi-unit buildings, the average of the points assigned to individual dwelling units may be used as the number of points awarded for this practice, rounded to the nearest whole number.)**	**4 for first compartment** **1 for each additional compartment in dwelling** **7 Max**
(2) All shower compartments in the dwelling unit(s) and common areas meet the requirements of 801.3(1) and all showerheads are in accordance with one of the following:	
(a) 2.0 to less than 2.5 gpm	**11 Additional**
(b) 1.6 to less than 2.0 gpm	**14 Additional**
(3) Any shower control that can shut off water flow without affecting temperature is installed. **(Points awarded per shower control.)**	**1** **3 Max**
For SI: 1 gallon per minute = 3.785 L/m	

GREEN BUILDING PRACTICES	POINTS
801.4 Lavatory faucets	

GREEN BUILDING PRACTICES	POINTS
801.4.1 Water-efficient lavatory faucets with a maximum flow rate of 1.5 gpm (5.68 L/m), tested at 60 psi (414 kPa) in accordance with ASME A112.18.1, are installed:	
(1) a bathroom (all faucets in a bathroom are in compliance)	1
(Points awarded for each bathroom. In multi-unit buildings, the average of the points assigned to individual dwelling units may be used as the number of points awarded for this practice, rounded to the nearest whole number.)	3 Max
(2) all lavatory faucets in the dwelling unit(s) and common areas	**6 Additional**

GREEN BUILDING PRACTICES	POINTS
801.4.2 Self-closing valve, motion sensor, metering, or pedal-activated faucet is installed to enable intermittent on/off operation. **(Points awarded per fixture.)**	1 3 Max

GREEN BUILDING PRACTICES	POINTS
801.5 Water closets and urinals. Water closets and urinals are in accordance with the following: **(Points awarded for 801.5(2) or 801.5(3), not both.)**	
(1) Gold and emerald levels: All water closets and urinals are in accordance with Section 801.5.	**Mandatory**
(2) A water closet is installed with an effective flush volume of 1.28 gallons (4.85 L) or less when tested in accordance with ASME A112.19.2/CSA B45.1 or ASME A112.19.14 as applicable, and is in accordance with EPA WaterSense *Tank-Type Toilets*. **(Points awarded per fixture. In multi-unit buildings, the average of the points assigned to individual dwelling units may be used as the number of points awarded for this practice, rounded to the nearest whole number.)**	**2** **6 Max**
(3) All water closets are in accordance with Section 801.5(2).	**11**
(4) All water closets are in accordance with Section 801.5(2) and one or more of the following are installed:	
(a) Water closets that have a flush volume of 1.2 gallons or less. **(Points awarded per toilet. In multi-unit buildings, the average of the points assigned to individual dwelling units may be used as the number of points awarded for this practice, rounded to the nearest whole number.)**	**1 Additional** **3 Additional Max**
(b) One or more urinals with a flush volume of 0.5 gallons (1.9L) or less when tested in accordance with ASME A112.19.2.	**1 Additional**
(c) One or more composting or waterless toilets and/or urinals.	**6 Additional**

801.6 Irrigation systems

801.6.1 Multi-stream, multi-trajectory rotating nozzles are installed in lieu of spray nozzles for turf or landscaping.	6

801.6.2 Drip irrigation is installed.	8 Max
(1) Drip irrigation is installed for landscape beds.	4
(2) Subsurface drip is installed for turf grass areas.	4

801.6.3 Landscape plan and implementation are executed by a certified WaterSense Professional or equivalent as approved by Adopting Entity.	5 Additional

801.6.4 Drip irrigation zones specifications show plant type by name and water use/need for each emitter. **(Points awarded only if specifications are implemented.)**	10 Additional

GREEN BUILDING PRACTICES	POINTS
801.6.5 The irrigation system(s) is controlled by a smart controller or no irrigation is installed. **(Points for 801.6.5(2) are not additive with points for 801.6.5(1).)**	
(1) Evapotranspiration (ET) based irrigation controller with a rain sensor or soil moisture sensor based irrigation controller.	8
(2) No irrigation is installed and a landscape plan is developed in accordance with Section 503.5, as applicable.	15

801.7 Rainwater collection and distribution. Rainwater collection and distribution is provided.	

801.7.1 Rainwater is used for irrigation in accordance with one of the following:	
(1) Rainwater is diverted for landscape irrigation without impermeable water storage	5
(2) Rainwater is diverted for landscape irrigation with impermeable water storage in accordance with one of the following:	
(a) 50 – 499 gallon storage capacity	5
(b) 500 – 2499 gallon storage capacity	10
(c) 2500 gallon or larger storage capacity (system is designed by a professional certified by The American Rainwater Catchment Systems Association or equivalent)	15
(d) All irrigation demands are met by rainwater capture (documentation demonstrating the water needs of the landscape is provided and the system is designed by a professional certified by The American Rainwater Catchment Systems Association or equivalent).	25

801.7.2 Rainwater is used for indoor domestic demand as follows. The system is designed by a professional certified by The American Rainwater Catchment Systems Association or equivalent.	
(1) Rainwater is used to supply an indoor appliance or fixture for any locally approved use.	5
(Points awarded per appliance or fixture.)	15 Max
(2) Rainwater provides for total domestic demand.	25

801.8 Sediment filters. Water filter is installed to reduce sediment and protect plumbing fixtures for the whole building or the entire dwelling unit.	1

GREEN BUILDING PRACTICES	POINTS

802
INNOVATIVE PRACTICES

802.1 Reclaimed, gray, or recycled water. Reclaimed, gray, or recycled water is used as permitted by applicable code. **(Points awarded for either Section 802.1(1) or 802.1(2), not both.)** **(Points awarded for either Section 802.5 or 802.1, not both.)**	
(1) each water closet flushed by reclaimed, gray, or recycled water	**5**
(Points awarded per fixture or appliance.)	**20 Max**
(2) irrigation from reclaimed, gray, or recycled water on-site	**10**

802.2 Automatic shutoff water devices. One of the following automatic shutoff water supply devices is installed. Where a fire sprinkler system is present, installer is to ensure the device will not interfere with the operation of the fire sprinkler system.	**2**
(1) excess water flow automatic shutoff	
(2) leak detection system with automatic shutoff	

802.3 Engineered biological system or intensive bioremediation system. An engineered biological system or intensive bioremediation system is installed and the treated water is used on site. Design and implementation are approved by appropriate regional authority.	**20**

802.4 Recirculating humidifier. Where a humidifier is required, a recirculating humidifier is used in lieu of a traditional "flow through" type.	**1**

802.5 Advanced wastewater treatment system. Advanced wastewater (aerobic) treatment system is installed and treated water is used on site. **(Points awarded for either Section 802.5 or 802.1, not both.)**	**20**

ICC 700-2012 NATIONAL GREEN BUILDING STANDARD™

CHAPTER 9

INDOOR ENVIRONMENTAL QUALITY

GREEN BUILDING PRACTICES	POINTS

901 POLLUTANT SOURCE CONTROL

901.0 Intent. Pollutant sources are controlled.	

901.1 Space and water heating options

901.1.1 Natural draft furnaces, boilers, or water heaters are not located in conditioned spaces, including conditioned crawlspaces, unless located in a mechanical room that has an outdoor air source and is sealed and insulated to separate it from the conditioned space(s). **(Points are awarded only for buildings that use natural draft combustion space or water heating equipment.)**	5
901.1.2 Air handling equipment or return ducts are not located in the garage, unless placed in isolated, air-sealed mechanical rooms with an outside air source.	5

901.1.3 The following combustion space heating or water heating equipment is installed within conditioned space:	
(1) all furnaces or all boilers	
(a) power vent furnace(s) or boiler(s)	3
(b) direct vent furnace(s) or boiler(s)	5
(2) all water heaters	
(a) power vent water heater(s)	3
(b) direct vent water heater(s)	5

901.1.4 Gas-fired fireplaces and direct heating equipment is listed and is installed in accordance with the NFPA 54, ICC IFGC, or the applicable local gas appliance installation code. Gas-fired fireplaces and direct heating equipment are vented to the outdoors.	**Mandatory**
901.1.5 Natural gas and propane fireplaces are direct vented, have permanently fixed glass fronts or gasketed doors, and comply with CSA Z21.88/CSA 2.33 or CSA Z21.50b/CSA 2.22b.	7

901.1.6 The following electric equipment is installed:	
(1) heat pump air handler in unconditioned space	2
(2) heat pump air handler in conditioned space	5

GREEN BUILDING PRACTICES	POINTS

901.2 Solid fuel-burning appliances

901.2.1 Solid fuel-burning fireplaces, inserts, stoves and heaters are code compliant and are in accordance with the following requirements:		**Mandatory**
(1)	Site-built masonry wood-burning fireplaces use outside combustion air and include a means of sealing the flue and the combustion air outlets to minimize interior air (heat) loss when not in operation.	**4**
(2)	Factory-built, wood-burning fireplaces are in accordance with the certification requirements of UL 127 and are EPA certified.	**6**
(3)	Wood stove and fireplace inserts, as defined in UL 1482 Section 3.8, are in accordance with the certification requirements of UL 1482 and are in accordance with the emission requirements of the EPA Certification and the State of Washington WAC 173-433-100(3).	**6**
(4)	Pellet (biomass) stoves and furnaces are in accordance with ASTM E1509 or are EPA certified.	**6**
(5)	Masonry heaters are in accordance with the definitions in ASTM E1602 and ICC IBC Section 2112.1.	**6**

901.2.2 Fireplaces, woodstoves, pellet stoves, or masonry heaters are not installed.	**7**

901.3 Garages. Garages are in accordance with the following:		
(1)	Attached garage	
	(a) Doors installed in the common wall between the attached garage and conditioned space are tightly sealed and gasketed.	**Mandatory 2**
	(b) A continuous air barrier is provided separating the garage space from the conditioned living spaces.	**Mandatory 2**
	(c) For one- and two-family dwelling units, a 100 cfm (47 L/s) or greater ducted or 70 cfm (33 L/s) cfm or greater unducted wall exhaust fan is installed and vented to the outdoors and is designed and installed for continuous operation or has controls (e.g., motion detectors, pressure switches) that activate operation for a minimum of 1 hour when either human passage door or roll-up automatic doors are operated. For ducted exhaust fans, the fan airflow rating and duct sizing are in accordance with Appendix A.	**8**
(2)	A carport is installed, the garage is detached from the building, or no garage is installed.	**10**

901.4 Wood materials. A minimum of 85 percent of material within a product group (i.e., wood structural panels, countertops, composite trim/doors, custom woodwork, and/or component closet shelving) is manufactured in accordance with the following:		**10 Max**
(1)	Structural plywood used for floor, wall, and/or roof sheathing is compliant with DOC PS 1 and/or DOC PS 2. OSB used for floor, wall, and/or roof sheathing is compliant with DOC PS 2. The panels are made with moisture-resistant adhesives. The trademark indicates these adhesives as follows: Exposure 1 or Exterior for plywood, and Exposure 1 for OSB.	**Mandatory**

GREEN BUILDING PRACTICES	POINTS
(2) Particleboard and MDF (medium density fiberboard) is manufactured and labeled in accordance with CPA A208.1 and CPA A208.2, respectively. **(Points awarded per product group.)**	2
(3) Hardwood plywood in accordance with HPVA HP-1. **(Points awarded per product group.)**	2
(4) Particleboard, MDF, or hardwood plywood is in accordance with CPA 4. **(Points awarded per product group.)**	3
(5) Composite wood or agrifiber panel products contain no added urea-formaldehyde or are in accordance with the CARB *Composite Wood Air Toxic Contaminant Measure Standard.* **(Points awarded per product group.)**	4
(6) Non-emitting products. **(Points awarded per product group.)**	4

901.5 Cabinets. A minimum of 85 percent of installed cabinets are in accordance with one or both of the following: **(Where both of the following practices are used, only 3 points are awarded.)**	
(1) All parts of the cabinet are made of solid wood or non-formaldehyde emitting materials such as metal or glass.	5
(2) The composite wood used in wood cabinets is in accordance with CARB *Composite Wood Air Toxic Contaminant Measure Standard* or equivalent as certified by a third-party program such as, but not limited to, those in Appendix D.	3

901.6 Carpets. Carpets are in accordance with the following:	
(1) Wall-to-wall carpeting is not installed adjacent to water closets and bathing fixtures.	**Mandatory**
(2) A minimum of 10 percent of the conditioned floor space has carpet and at least 85 percent of installed carpet area and/or carpet cushion (padding) are in accordance with the emission levels of CDPH/EHLB Standard Method v1.1 except footnote b in Table 4.1 does not apply (i.e., allowable maximum formaldehyde concentration is 16.5 µg/m^3 (13.5 ppb)). Product is tested by a laboratory with the CDPH/EHLB Standard Method v1.1 within the laboratory scope of accreditation to ISO/IEC 17025 and certified by a third-party program accredited to ISO Guide 65, such as, but not limited to, those in Appendix D.	
(a) carpet	6
(b) carpet cushion	2

901.7 Hard-surface flooring. A minimum of 10 percent of the conditioned floor space has pre-finished hard-surface flooring installed and a minimum of 85 percent of all prefinished installed hard-surface flooring is in accordance with the emission concentration limits of CDPH/EHLB Standard Method v1.1 except footnote b in Table 4.1 does not apply (i.e., allowable maximum formaldehyde concentration is 16.5 µg/m^3 (13.5 ppb)). Emission levels are determined by a laboratory accredited to ISO/IEC 17025 and the CDPH/EHLB Standard Method v1.1 is in its scope of accreditation. The product is certified by a third-party program accredited to ISO Guide 65, such as, but not limited to, those found in Appendix D.	6

GREEN BUILDING PRACTICES	POINTS

Where post-manufacture coatings or surface applications have not been applied, the following hard surface flooring types are deemed to comply with the emission requirements of this practice:

(a)	Ceramic tile flooring
(b)	Organic-free, mineral-based flooring
(c)	Clay masonry flooring
(d)	Concrete masonry flooring
(e)	Concrete flooring
(f)	Metal flooring

901.8 Wall coverings. A minimum of 10 percent of the interior wall surfaces are covered and a minimum of 85 percent of wall coverings are in accordance with the emission concentration limits of CDPH/EHLB Standard Method v1.1 except footnote b in Table 4.1 does not apply (i.e., allowable maximum formaldehyde concentration is 16.5 $\mu g/m^3$ (13.5 ppb)). Emission levels are determined by a laboratory accredited to ISO/IEC 17025 and the CDPH/EHLB Standard Method v1.1 is in its scope. The product is certified by a third-party program accredited to ISO Guide 65, such as, but not limited to, those in Appendix D.	4

901.9 Interior architectural coatings. A minimum of 85 percent of the interior architectural coatings are in accordance with either Section 901.9.1 or Section 901.9.3, not both. A minimum of 85 percent of architectural colorants are in accordance with Section 901.9.2.

901.9.1 Site-applied interior architectural coatings, which are inside the water proofing envelope, are in accordance with one or more of the following:	5

(1)	Zero VOC as determined by EPA Method 24 (VOC content is below the detection limit for the method)
(2)	GreenSeal GS-11
(3)	CARB *Suggested Control Measure for Architectural Coatings* (see Table 901.9.1).

Table 901.9.1
VOC Content Limits For Architectural Coatings[a,b,c]

Coating Category	LIMIT[d] (g/l)
Flat Coatings	50
Non-flat Coatings	100
Non-flat High-Gloss Coatings	150
Specialty Coatings:	
Aluminum Roof Coatings	400
Basement Specialty Coatings	400
Bituminous Roof Coatings	50
Bituminous Roof Primers	350
Bond Breakers	350
Concrete Curing Compounds	350
Concrete/Masonry Sealers	100
Driveway Sealers	50
Dry Fog Coatings	150

GREEN BUILDING PRACTICES	POINTS

Coating Category	LIMIT[d] (g/l)
Faux Finishing Coatings	350
Fire Resistive Coatings	350
Floor Coatings	100
Form-Release Compounds	250
Graphic Arts Coatings (Sign Paints)	500
High Temperature Coatings	420
Industrial Maintenance Coatings	250
Low Solids Coatings	120[e]
Magnesite Cement Coatings	450
Mastic Texture Coatings	100
Metallic Pigmented Coatings	500
Multi-Color Coatings	250
Pre-Treatment Wash Primers	420
Primers, Sealers, and Undercoaters	100
Reactive Penetrating Sealers	350
Recycled Coatings	250
Roof Coatings	50
Rust Preventative Coatings	250
Shellacs, Clear	730
Shellacs, Opaque	550
Specialty Primers, Sealers, and Undercoaters	100
Stains	250
Stone Consolidants	450
Swimming Pool Coatings	340
Traffic Marking Coatings	100
Tub and Tile Refinish Coatings	420
Waterproofing Membranes	250
Wood Coatings	275
Wood Preservatives	350
Zinc-Rich Primers	340

a. The specified limits remain in effect unless revised limits are listed in subsequent columns in the table.
b. Values in this table are derived from those specified by the California Air Resources Board, Architectural Coatings Suggested Control Measure, February 1, 2008.
c. Table 901.9.1 architectural coating regulatory category and VOC content compliance determination shall conform to the California Air Resources Board Suggested Control Measure for Architectural Coatings dated February 1, 2008.
d. Limits are expressed as VOC Regulatory (except as noted), thinned to the manufacturer's maximum thinning recommendation, excluding any colorant added to tint bases.
e. Limit is expressed as VOC actual.

GREEN BUILDING PRACTICES	POINTS
901.9.2 Architectural coating colorant additive VOC content is in accordance with Table 901.9.2. **(Points for 901.9.2 are awarded only if base architectural coating is in accordance with 901.9.1.)**	1

Table 901.9.2
VOC Content Limits for Colorants

Colorant	LIMIT (g/l)
Architectural Coatings, excluding IM Coatings	50
Solvent-Based IM	600
Waterborne IM	50

GREEN BUILDING PRACTICES	POINTS
901.9.3 Site-applied interior architectural coatings, which are inside the waterproofing envelope, are in accordance with the emission levels of CDPH/EHLB Standard Method v1.1 , except footnote b in Table 4.1 does not apply (i.e., allowable maximum formaldehyde concentration is 16.5 µg/m^3 (13.5 ppb)). Emission levels are determined by a laboratory accredited to ISO/IEC 17025 and the CDPH/EHLB Standard Method v1.1 in its scope of accreditation. The product is certified by a third-party program accredited to ISO Guide 65, such as, but not limited to, those found in Appendix D.	8

	GREEN BUILDING PRACTICES	POINTS
	901.10 Interior adhesives and sealants. A minimum of 85 percent of site-applied adhesives and sealants located inside the waterproofing envelope are in accordance with one of the following, as applicable.	
(1)	The emission levels are in accordance with CDPH/EHLB Standard Method v1.1 except footnote b in Table 4.1 does not apply (i.e., allowable maximum formaldehyde concentration is 16.5 µg/m^3 (13.5 ppb)). Emission levels are determined by a laboratory accredited to ISO/IEC 17025 and the CDPH/EHLB Standard Method v1.1 is in its scope of accreditation. The product is certified by a third-party program accredited to ISO Guide 65, such as, but not limited to, those found in Appendix D.	8
(2)	GreenSeal GS-36.	5
(3)	SCAQMD Rule 1168 in accordance with Table 901.10(3), excluding products that are sold in 16 ounce containers or less and are regulated by the California Air Resources Board (CARB) Consumer Products Regulations.	5

Table 901.10(3)
Site Applied Adhesive and Sealants VOC Limits[a,b]

ADHESIVE OR SEALANT	VOC LIMIT (g/l)
Indoor carpet adhesives	50
Carpet pad adhesives	50
Outdoor carpet adhesives	150
Wood flooring adhesive	100
Rubber floor adhesives	60
Subfloor adhesives	50
Ceramic tile adhesives	65
VCT and asphalt tile adhesives	50
Drywall and panel adhesives	50
Cove base adhesives	50

GREEN BUILDING PRACTICES	POINTS

ADHESIVE OR SEALANT	VOC LIMIT (g/l)
Multipurpose construction adhesives	70
Structural glazing adhesives	100
Single ply roof membrane adhesives	250
Architectural sealants	250
Architectural sealant primer Non-porous Porous	 250 775
Modified bituminous sealant primer	500
Other sealant primers	750
CPVC solvent cement	490
PVC solvent cement	510
ABS solvent cement	325
Plastic cement welding	250
Adhesive primer for plastic	550
Contact adhesive	80
Special purpose contact adhesive	250
Structural wood member adhesive	140

a. VOC limit less water and less exempt compounds in grams/liter
b. For low-solid adhesives and sealants, the VOC limit is expressed in grams/liter of material as specified in Rule 1168. For all other adhesives and sealants, the VOC limits are expressed as grams of VOC per liter of adhesive or sealant less water and less exempt compounds as specified in Rule 1168.

901.11 Insulation. Emissions of 85 percent of wall, ceiling, and floor insulation materials are in accordance with the emission levels of CDPH/EHLB Standard Method v1.1 except footnote b in Table 4.1 does, not apply (i.e., allowable maximum formaldehyde concentration is 16.5 $\mu g/m^3$ (13.5 ppb)). Emission levels are determined by a laboratory accredited to ISO/IEC 17025 and the CDPH/EHLB Standard Method v1.1 is in its scope of accreditation. Insulation is certified by a third-party program accredited to ISO Guide 65, such as, but not limited to, those in Appendix D.	4
901.12 Carbon monoxide (CO) alarms. Where not required by local codes, a carbon monoxide (CO) alarm is installed in a central location outside of each separate sleeping area in the immediate vicinity of the bedrooms. The CO alarm(s) is located in accordance with NFPA 720 and is hardwired with a battery backup. The alarm device(s) is certified by a third-party for conformance to either CSA 6.19 or UL 2034.	4

901.13 Building entrance pollutants control. Pollutants are controlled at all main building entrances by one of the following methods:	
(1) Exterior grilles or mats are installed in a fixed manner and may be removable for cleaning.	1
(2) Interior grilles or mats are installed in a fixed manner and may be removable for cleaning.	1

GREEN BUILDING PRACTICES	POINTS

901.14 Non-smoking areas. Environmental tobacco smoke is minimized by one or more of the following:	
(1) All interior common areas of a multi-unit building are designated as non-smoking areas with posted signage.	1
(2) Exterior smoking areas of a multi-unit building are designated with posted signage and located a minimum of 25 feet from entries, outdoor air intakes, and operable windows.	1

902
POLLUTANT CONTROL

902.0 Intent. Pollutants generated in the building are controlled.

902.1 Spot ventilation.

902.1.1 Spot ventilation is in accordance with the following:	
(1) Bathrooms are vented to the outdoors. The minimum ventilation rate is 50 cfm (23.6 L/s) for intermittent operation or 20 cfm (9.4 L/s) for continuous operation in bathrooms. **(Points are awarded only if a window complying with IRC Section R303.3 is provided in addition to mechanical ventilation.)**	**Mandatory** 1
(2) Clothes dryers are vented to the outdoors.	**Mandatory**
(3) Kitchen exhaust units and/or range hoods are ducted to the outdoors and have a minimum ventilation rate of 100 cfm (47.2 L/s) for intermittent operation or 25 cfm (11.8 L/s) for continuous operation.	8

902.1.2 Bathroom and/or laundry exhaust fan is provided with an automatic timer and/or humidistat:	**11 Max**
(1) for first device	5
(2) for each additional device	2

902.1.3 Kitchen range, bathroom, and laundry exhaust are verified to air flow specification. Ventilation airflow at the point of exhaust is tested to a minimum of:	8
(a) 100 cfm (47.2 L/s) intermittent or 25 cfm (11.8 L/s) continuous for kitchens, and	
(b) 50 cfm (23.6 L/s) intermittent or 20 cfm (9.4 L/s) continuous for bathrooms and/or laundry	

902.1.4 Exhaust fans are ENERGY STAR, as applicable.	**12 Max**
(1) ENERGY STAR, or equivalent, fans **(Points awarded per fan.)**	2
(2) ENERGY STAR, or equivalent, fans operating at or below 1 sone **(Points awarded per fan.)**	3

GREEN BUILDING PRACTICES	POINTS

902.2 Building ventilation systems.

902.2.1 One of the following whole building ventilation systems is implemented and is in accordance with the specifications of Appendix B.	Mandatory where the maximum air infiltration rate is less than 5 ACH50
(1) exhaust or supply fan(s) ready for continuous operation and with appropriately labeled controls	3
(2) balanced exhaust and supply fans with supply intakes located in accordance with the manufacturer's guidelines so as to not introduce polluted air back into the building	6
(3) heat-recovery ventilator	7
(4) energy-recovery ventilator	8

902.2.2 Ventilation airflow is tested to achieve the design fan airflow at point of exhaust in accordance with Section 902.2.1.	4

902.2.3 MERV filters 8 or greater are installed on central forced air systems and are accessible. Designer or installer is to verify that the HVAC equipment is able to accommodate the greater pressure drop of MERV 8 filters.	3

902.3 Radon control. Radon control measures are in accordance with ICC IRC Appendix F. Zones as defined in Figure 9(1).	
(1) Buildings located in Zone 1	Mandatory
(a) a passive radon system is installed	7
(b) an active radon system is installed	10
(2) Buildings located in Zone 2 or Zone 3	
(a) a passive or active radon system is installed	7

902.4 HVAC system protection. One of the following HVAC system protection measures is performed.	3
(1) HVAC supply registers (boots), return grilles, and rough-ins are covered during construction activities to prevent dust and other pollutants from entering the system.	
(2) Prior to owner occupancy, HVAC supply registers (boots), return grilles, and duct terminations are inspected and vacuumed. In addition, the coils are inspected and cleaned and the filter is replaced if necessary.	

902.5 Central vacuum systems. Central vacuum system is installed and vented to the outside.	3

902.6 Living space contaminants. The living space is sealed in accordance with Section 701.4.3.1 to prevent unwanted contaminants.	Mandatory

GREEN BUILDING PRACTICES	POINTS

903
MOISTURE MANAGEMENT: VAPOR, RAINWATER, PLUMBING, HVAC

903.0 Intent. Moisture and moisture effects are controlled.	

903.1 Plumbing	

903.1.1 Cold water pipes in unconditioned spaces are insulated to a minimum of R-4 with pipe insulation or other covering that adequately prevents condensation.	2

903.1.2 Plumbing is not installed in unconditioned spaces.	5

903.2 Duct insulation. Ducts are in accordance with one of the following.	
(1) All HVAC ducts, plenums, and trunks are located in conditioned space.	1
(2) All HVAC ducts, plenums, and trunks are in conditioned space. All HVAC ducts are insulated to a minimum of R4.	3

903.3 Relative humidity. In climate zones 1A, 2A, 3A, 4A, and 5A as defined by Figure 6(1), equipment is installed to maintain relative humidity (RH) at or below 60 percent using one of the following: **(Points not awarded in other climate zones.)**	7
(1) additional dehumidification system(s)	
(2) central HVAC system equipped with additional controls to operate in dehumidification mode	

904
INNOVATIVE PRACTICES

904.1 Humidity monitoring system. A humidity monitoring system is installed with a mobile base unit that displays readings of temperature and relative humidity. The system has a minimum of two remote sensor units. One remote sensor unit is placed permanently inside the conditioned space in a central location, excluding attachment to exterior walls, and another remote sensor unit is placed permanently outside of the conditioned space.	2

904.2 Kitchen exhaust. A kitchen exhaust unit(s) that equals or exceeds 400 cfm (189 L/s) is installed, and makeup air is provided.	2

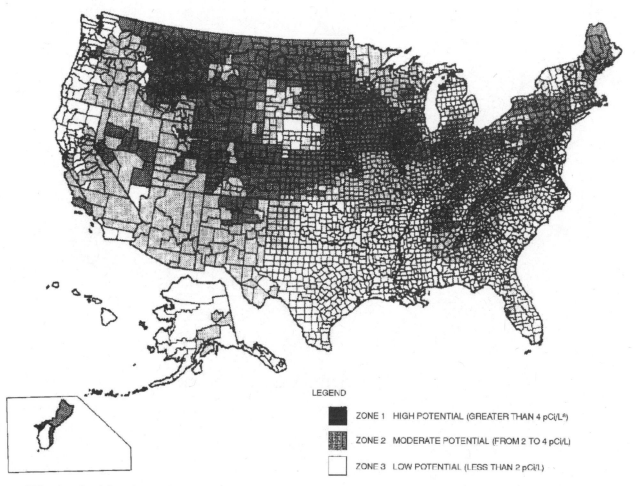

LEGEND

ZONE 1   HIGH POTENTIAL (GREATER THAN 4 pCi/L[a])

ZONE 2   MODERATE POTENTIAL (FROM 2 TO 4 pCi/L)

ZONE 3   LOW POTENTIAL (LESS THAN 2 pCi/L)

a. pCi/L standard for picocuries per liter of radon gas. EPA recommends that all homes that measure 4 pCi/L and greater be mitigated.

The United States Environmental Protection Agency and the United States Geological Survey have evaluated the radon potential in the United States and have developed a map of radon zones designed to assist building officials in deciding whether radon-resistant features are applicable in new construction.

The map assigns each of the 3,141 counties in the United States to one of three zones based on radon potential. Each zone designation reflects the average short-term radon measurement that can be expected to be measured in a building without the implementation of radon control methods. The radon zone designation of highest priority is Zone 1. This table lists the Zone 1 counties illustrated on the map. More detailed information can be obtained from state-specific booklets (EPA-402-R-93-021 through 070) available through state radon offices or from U.S. EPA regional offices.

FIGURE 9(1)
EPA MAP OF RADON ZONES

THIS PAGE INTENTIONALLY LEFT BLANK

CHAPTER 10

OPERATION, MAINTENANCE, AND BUILDING OWNER EDUCATION

GREEN BUILDING PRACTICES	POINTS

1001
BUILDING OWNERS' MANUAL FOR ONE- AND TWO-FAMILY DWELLINGS

1001.0 Intent. Information on the building's use, maintenance, and green components is provided.

1001.1 A building owner's manual is provided that includes the following, as available and applicable. **(Points awarded per two items. Points awarded for both mandatory and non-mandatory items.)**		1 8 Max
(1)	A green building program certificate or completion document.	Mandatory
(2)	List of green building features (can include the national green building checklist).	Mandatory
(3)	Product manufacturer's manuals or product data sheet for installed major equipment, fixtures, and appliances. If product data sheet is in the building owners' manual, manufacturer's manual may be attached to the appliance in lieu of inclusion in the building owners' manual.	Mandatory
(4)	Maintenance checklist.	
(5)	Information on local recycling programs.	
(6)	Information on available local utility programs that purchase a portion of energy from renewable energy providers.	
(7)	Explanation of the benefits of using energy-efficient lighting systems [e.g., compact fluorescent light bulbs, light emitting diode (LED)] in high-usage areas.	
(8)	A list of practices to conserve water and energy.	
(9)	Local public transportation options.	
(10)	A diagram showing the location of safety valves and controls for major building systems.	
(11)	Where frost-protected shallow foundations are used, owner is informed of precautions including:	
	(a) instructions to not remove or damage insulation when modifying landscaping.	
	(b) providing heat to the building as required by the ICC IRC or IBC.	
	(c) keeping base materials beneath and around the building free from moisture caused by broken water pipes or other water sources.	
(12)	A list of local service providers that offer regularly scheduled service and maintenance contracts to ensure proper performance of equipment and the structure (e.g., HVAC, water-heating equipment, sealants, caulks, gutter and downspout system, shower and/or tub surrounds, irrigation system).	

GREEN BUILDING PRACTICES	POINTS
(13) A photo record of framing with utilities installed. Photos are taken prior to installing insulation, clearly labeled, and included as part of the building owners' manual.	
(14) List of common hazardous materials often used around the building and instructions for proper handling and disposal of these materials.	
(15) Information on organic pest control, fertilizers, deicers, and cleaning products.	
(16) Information on native landscape materials and/or those that have low water requirements.	
(17) Information on methods of maintaining the building's relative humidity in the range of 30 percent to 60 percent.	
(18) Instructions for inspecting the building for termite infestation.	
(19) Instructions for maintaining gutters and downspouts and importance of diverting water a minimum of 5 feet away from foundation.	
(20) A narrative detailing the importance of maintenance and operation in retaining the attributes of a green-built building.	
(21) Where stormwater management measures are installed on the lot, information on the location, purpose, and upkeep of these measures.	

1002
TRAINING OF BUILDING OWNERS ON OPERATION AND MAINTENANCE FOR ONE- AND TWO-FAMILY DWELLINGS AND MULTI-UNIT BUILDINGS

1002.1 Training of building owners. Building owners are familiarized with the role of occupants in achieving green goals. On-site training is provided to the responsible party(ies) regarding equipment operation and maintenance, control systems, and occupant actions that will improve the environmental performance of the building. These include:	8
(1) HVAC filters	
(2) thermostat operation and programming	
(3) lighting controls	
(4) appliances operation	
(5) water heater settings and hot water use	
(6) fan controls	
(7) recycling practices	

1003
CONSTRUCTION, OPERATION, AND MAINTENANCE MANUALS AND TRAINING FOR MULTI-UNIT BUILDINGS

1003.0 Intent. Manuals are provided to the responsible parties (owner, management, tenant, and/or maintenance team) regarding the construction, operation, and maintenance of the building. Paper or digital format manuals are to include information regarding those aspects of the building's construction, maintenance, and operation that are within the area of responsibilities of the respective recipient. One or more responsible parties are to receive a copy of all documentation for archival purposes.

GREEN BUILDING PRACTICES	POINTS
1003.1 Building construction manual. A building construction manual, including five or more of the following, is compiled and distributed in accordance with Section 1003.0. **(Points awarded per two items. Points awarded for both mandatory and non-mandatory items.)**	1
(1) A narrative detailing the importance of constructing a green building, including a list of green building attributes included in the building. This narrative is included in all responsible parties' manuals.	Mandatory
(2) A local green building program certificate as well as a copy of the *National Green Building Standard*[TM], as adopted by the Adopting Entity, and the individual measures achieved by the building.	Mandatory
(3) Warranty, operation, and maintenance instructions for all equipment, fixtures, appliances, and finishes.	Mandatory
(4) Record drawings of the building.	
(5) A record drawing of the site including stormwater management plans, utility lines, landscaping with common name and genus/species of plantings.	
(6) A diagram showing the location of safety valves and controls for major building systems.	
(7) A list of the type and wattage of light bulbs installed in light fixtures.	
(8) A photo record of framing with utilities installed. Photos are taken prior to installing insulation and clearly labeled.	

GREEN BUILDING PRACTICES	POINTS
1003.2 Operations manual. Operations manuals are created and distributed to the responsible parties in accordance with Section 1003.0. Between all of the operation manuals, five or more of the following options are included. **(Points awarded per two items. Points awarded for both mandatory and non-mandatory items.)**	1
(1) A narrative detailing the importance of operating and living in a green building. This narrative is included in all responsible parties' manuals.	Mandatory
(2) A list of practices to conserve water and energy (e.g., turning off lights when not in use, switching the rotation of ceiling fans in changing seasons, purchasing ENERGY STAR appliances and electronics).	Mandatory
(3) Information on methods of maintaining the building's relative humidity in the range of 30 percent to 60 percent.	
(4) Information on opportunities to purchase renewable energy from local utilities or national green power providers and information on utility and tax incentives for the installation of on-site renewable energy systems.	
(5) Information on local and on-site recycling and hazardous waste disposal programs and, if applicable, building recycling and hazardous waste handling and disposal procedures.	
(6) Local public transportation options.	
(7) Explanation of the benefits of using compact fluorescent light bulbs, LEDs, or other high-efficiency lighting.	
(8) Information on native landscape materials and/or those that have low water requirements.	
(9) Information on the radon mitigation system, where applicable.	

GREEN BUILDING PRACTICES	POINTS

(10) A procedure for educating tenants in rental properties on the proper use, benefits, and maintenance of green building systems including a maintenance staff notification process for improperly functioning equipment.	

1003.3 Maintenance manual. Maintenance manuals are created and distributed to the responsible parties in accordance with Section 1003.0. Between all of the maintenance manuals, five or more of the following options are included. **(Points awarded per two items. Points awarded for both mandatory and non-mandatory items.)**	**1**
(1) A narrative detailing the importance of maintaining a green building. This narrative is included in all responsible parties' manuals.	**Mandatory**
(2) A list of local service providers that offer regularly scheduled service and maintenance contracts to ensure proper performance of equipment and the structure (e.g., HVAC, water-heating equipment, sealants, caulks, gutter and downspout system, shower and/or tub surrounds, irrigation system).	
(3) User-friendly maintenance checklist that includes:	
(a) HVAC filters	
(b) thermostat operation and programming	
(c) lighting controls	
(d) appliances and settings	
(e) water heater settings	
(f) fan controls	
(4) List of common hazardous materials often used around the building and instructions for proper handling and disposal of these materials.	
(5) Information on organic pest control, fertilizers, deicers, and cleaning products.	
(6) Instructions for maintaining gutters and downspouts and the importance of diverting water a minimum of 5 feet away from foundation.	
(7) Instructions for inspecting the building for termite infestation.	
(8) A procedure for rental tenant occupancy turnover that preserves the green features.	
(9) An outline of a formal green building training program for maintenance staff.	

1004
INNOVATIVE PRACTICES

1004.1 (Reserved)

CHAPTER 11

REMODELING

GREEN BUILDING PRACTICES	POINTS

11.500
LOT DESIGN, PREPARATION, AND DEVELOPMENT

11.500.0 Intent. This section applies to the lot and changes to the lot due to remodeling of an existing building.

11.501
LOT SELECTION

11.501.2 Multi-modal transportation. A range of multi-modal transportation choices are promoted by one or more of the following:	
(1) The building is located within one-half mile (805 m) of pedestrian access to a mass transit system or within five miles (8046 m) of a mass transit station with provisions for parking.	4
(2) The building is located within one-half mile (805 m) of six or more community resources (e.g., recreational facilities (such as pools, tennis courts, basketball courts), parks, grocery store, post office, place of worship, community center, daycare center, bank, school, restaurant, medical/dental office, Laundromat/dry cleaner).	4
(3) The building is on a lot located within a community that has rights-of-way specifically dedicated to bicycle use in the form of paved paths or bicycle lanes, or is on an infill lot located within 1/2 mile of a bicycle lane designated by the jurisdiction.	5

11.502
PROJECT TEAM, MISSION STATEMENT, AND GOALS

11.502.1 Project team, mission statement, and goals. A knowledgeable team is established and team member roles are identified with respect to green lot design, preparation, and development. The project's green goals and objectives are written into a mission statement.	4

11.503
LOT DESIGN

11.503.0 Intent. The lot is designed to avoid detrimental environmental impacts first, to minimize any unavoidable impacts, and to mitigate for those impacts that do occur. The project is designed to minimize environmental impacts and to protect, restore, and enhance the natural features and environmental quality of the lot.

**(Points awarded only if
the intent of the design is implemented.)**

GREEN BUILDING PRACTICES	POINTS

11.503.1 Natural resources. Natural resources are conserved by one or more of the following:

(1)	A natural resources inventory is completed under the direction of a qualified professional.	5
(2)	A plan is implemented to conserve the elements identified by the resource inventory as high-priority resources.	6
(3)	Items listed for protection in the resource inventory plan are protected under the direction of a qualified professional.	4
(4)	Basic training in tree or other natural resource protection is provided for the on-site supervisor.	4
(5)	All tree pruning on-site is conducted by a Certified Arborist.	3
(6)	Ongoing maintenance of vegetation on the lot during construction is in accordance with TCIA A300 or locally accepted best practices.	4
(7)	Where a lot adjoins a landscaped common area, a protection plan from the remodeling construction activities next to the common area is implemented.	5

11.503.2 Slope disturbance. Slope disturbance is minimized by one or more of the following:

(1)	The use of terrain-adaptive architecture including terracing, retaining walls, landscaping, or other re-stabilization techniques.	5
(2)	Hydrological/soil stability study is completed and used to guide the design of any additions to buildings on the lot.	4
(3)	All or a percentage of new driveways and parking are aligned with natural topography to reduce cut and fill.	
	(a) 10 percent to 25 percent	3
	(b) 25 percent to 75 percent	4
	(c) greater than 75 percent	6
(4)	Long-term erosion effects are reduced through the design and implementation of terracing, retaining walls, landscaping, or restabilization techniques.	5
(5)	Underground parking uses the natural slope for parking entrances.	5

11.503.3 Soil disturbance and erosion. Soil disturbance and erosion are minimized by one or more of the following: (also see Section 11.504.3)

(1)	Remodeling construction activities are scheduled to minimize length of time that soils are exposed.	5
(2)	The new utilities on the lot are designed to use one or more alternative means:	5
	(a) tunneling instead of trenching	
	(b) use of smaller (low ground pressure) equipment or geomats to spread the weight of construction equipment	
	(c) shared utility trenches or easements	
	(d) placement of utilities under paved surfaces instead of yards	
(3)	Limits of new clearing and grading are demarcated on the lot plan.	5

GREEN BUILDING PRACTICES	POINTS

11.503.4 Stormwater management. Stormwater management includes one or more of the following low-impact development techniques:		
(1)	Natural water and drainage features are preserved and used.	6
(2)	Facilities that minimize concentrated flows and simulate flows found in natural hydrology by the use of vegetative swales, french drains, wetlands, drywells, rain gardens, and similar infiltration features.	7
(3)	All or a percentage of impervious surfaces are minimized and permeable materials are used for driveways, parking areas, walkways, and patios.	
	(a) less than 25 percent	2
	(b) 25 percent to 75 percent	4
	(c) greater than 75 percent	6
(4)	A minimum of 50 percent of the roof is vegetated (green roof) using technology capable of withstanding the climate conditions of the jurisdiction and the microclimate conditions of the building site. Invasive plant species are not permitted.	5
(5)	Stormwater management practices that manage rainfall on-site and prevent the off-site discharge from all storms up to and including the volume of the 95th percentile storm event.	6

11.503.5 Landscape plan. A landscape plan for the lot is developed to limit water and energy use while preserving or enhancing the natural environment.		
(Where "front" only or "rear" only plan is implemented, only half of the points (rounding down to a whole number) are awarded for Items (1)-(6)		
(1)	Where a lot is less than 50 percent turf, a plan is formulated to restore or enhance natural vegetation that is cleared during construction. Landscaping is phased to coincide with achievement of final grades to ensure denuded areas are quickly vegetated.	6
(2)	Turf grass species, other vegetation, and trees that are native or regionally appropriate for local growing conditions are selected and specified on the lot plan.	4
(3)	The percentage of turf areas that is designed to be mowed is limited and shown on the lot plan. The percentage is based on the landscaped area of the lot not including the home footprint, hardscape, and any undisturbed natural areas.	
	(a) 0 percent or EPA WaterSense Water Budget Tool is used to determine the maximum percentage of turf areas	5
	(b) greater than 0 percent to less than 20 percent	4
	(c) 20 percent to less than 40 percent	3
	(d) 40 percent to 60 percent	2
(4)	Plants with similar watering needs are grouped (hydrozoning) and shown on the lot plan.	5
(5)	Summer shading by planting installed to shade a minimum of 30 percent of building walls. To conform to summer shading, the effective shade coverage (five years after planting) is the arithmetic mean of the shade coverage calculated at 10 am for eastward facing walls, noon for southward facing walls, and 3 pm for westward facing walls on the summer solstice.	5

GREEN BUILDING PRACTICES	POINTS
(6) Vegetative wind breaks or channels are designed to protect the lot and immediate surrounding lots as appropriate for local conditions.	4
(7) Site- or community-generated tree trimmings or stump grinding of regionally appropriate trees are used on the site to provide protective mulch during construction or for landscaping.	3
(8) An integrated pest management plan is developed to minimize chemical use in pesticides and fertilizers.	4

11.503.6 Wildlife habitat. Measures are planned to support wildlife habitat and include at least two of the following:	
(1) Plants and gardens that encourage wildlife, such as bird and butterfly gardens.	3
(2) Inclusion of a certified "backyard wildlife" program.	3
(3) The lot is adjacent to a wildlife corridor, fish and game park, or preserved areas and is designed with regard for this relationship.	3
(4) Outdoor lighting techniques are utilized with regard for wildlife.	3

11.503.7 Environmentally sensitive areas. The lot is in accordance with one or both of the following.	
(1) The lot does not contain any environmentally sensitive areas that are disturbed during remodeling.	4
(2) Environmentally sensitive areas compromised during remodeling are mitigated or restored.	4

11.504
LOT CONSTRUCTION

11.504.0 Intent. Environmental impact during construction is avoided to the extent possible; impacts that do occur are minimized, and any significant impacts are mitigated.

11.504.1 On-site supervision and coordination. On-site supervision and coordination is provided during on-lot-lot clearing, grading, trenching, paving, and installation of utilities to ensure that specified green development practices are implemented. (also see Section 11.503.3)	4

11.504.2 Trees and vegetation. Designated trees and vegetation are preserved by one or more of the following:	
(1) Fencing or equivalent is installed to protect trees and other vegetation.	3
(2) Trenching, significant changes in grade, and compaction of soil and critical root zones in all "tree save" areas as shown on the lot plan are avoided.	5
(3) Damage to designated existing trees and vegetation is mitigated during construction through pruning, root pruning, fertilizing, and watering.	4

GREEN BUILDING PRACTICES	POINTS

11.504.3 Soil disturbance and erosion implementation. On-site soil disturbance and erosion during remodeling are minimized by one or more of the following in accordance with the SWPPP or applicable plan: (also see Section 11.503.3)

(1)	Sediment and erosion controls are installed on the lot and maintained in accordance with the stormwater pollution prevention plan, where required.	5
(2)	Limits of clearing and grading are staked out on the lot.	5
(3)	"No disturbance" zones are created using fencing or flagging to protect vegetation and sensitive areas on the lot from construction activity.	5
(4)	Topsoil from either the lot or the site development is stockpiled and stabilized for later use and used to establish landscape plantings on the lot.	5
(5)	Soil compaction from construction equipment is reduced by distributing the weight of the equipment over a larger area (laying lightweight geogrids, mulch, chipped wood, plywood, OSB, metal plates, or other materials capable of weight distribution in the pathway of the equipment).	4
(6)	Disturbed areas on the lot that are complete or to be left unworked for 21 days or more are stabilized within 14 days using methods as recommended by the EPA, or in the approved SWPPP, where required.	3
(7)	Soil is improved with organic amendments and mulch.	3
(8)	Newly installed utilities on the lot are installed using one or more alternative means (e.g., tunneling instead of trenching, use of smaller equipment, use of low ground pressure equipment, use of geomats, shared utility trenches or easements).	5

11.505
INNOVATIVE PRACTICES

11.505.0 Intent. Innovative lot design, preparation and development practices are used to enhance environmental performance. Waivers or variances from local development regulations are obtained, and innovative zoning is used to implement such practices.

11.505.1 Driveways and parking areas. Driveways and parking areas are minimized by one or more of the following:

(1)	Off-street parking areas are shared or driveways are shared. Waivers or variances from local development regulations are obtained to implement such practices, if required.	5
(2)	In a multi-unit project, parking capacity does not exceed the local minimum requirements.	5
(3)	Structured parking is utilized to reduce the footprint of surface parking areas.	
	(a) 25 percent to less than 50 percent	4
	(b) 50 percent to 75 percent	5
	(c) greater than 75 percent	6

GREEN BUILDING PRACTICES	POINTS
11.505.2 Heat island mitigation. Heat island effect is mitigated by one or both of the following.	4
(1) Hardscape: Not less than 50 percent of the surface area of the hardscape on the lot meets one or a combination of the following methods.	5
(a) Shading of hardscaping: Shade is provided from existing or new vegetation (within five years) or from trellises. Shade of hardscaping is to be measured on the summer solstice at noon.	
(b) Light-colored hardscaping: Horizontal hardscaping materials are installed with a solar reflectance index (SRI) of 29 or greater. The SRI is calculated in accordance with ASTM E1980. A default SRI value of 35 for new concrete without added color pigment is permitted to be used instead of measurements.	
(c) Permeable hardscaping: Permeable hardscaping materials are installed.	
(2) Roofs: Not less than 75 percent of the exposed surface of the roof is in accordance with one or a combination of the following methods.	5
(a) Minimum initial SRI of 78 for a low-sloped roof (a slope less than or equal to 2:12) and a minimum initial SRI of 29 for a steep-sloped roof (a slope of more than 2:12). The SRI is calculated in accordance with ASTM E1980. Roof products shall be certified and labeled.	
(b) Roof is vegetated using technology capable of withstanding the climate conditions of the jurisdiction and the microclimate conditions of the building lot. Invasive plant species are not permitted.	

11.505.3 Density. The average density on the lot on a net developable area basis is:	
(1) 7 to less than 14 dwelling units per acre (per 4,047 m^2)	5
(2) 14 to less than 21 dwelling units per acre (per 4,047 m^2)	8
(3) 21 or greater dwelling units per acre (per 4,047 m^2)	11

11.505.4 Mixed-use development. The lot contains a mixed-use building.	8

11.505.5 Community Garden(s). A portion of the lot is established as a community garden(s), available to residents of the lot, to provide for local food production to residents or area consumers.	3

GREEN BUILDING PRACTICES	POINTS

11.601
QUALITY OF CONSTRUCTION MATERIALS AND WASTE

11.601.0 Intent. Design and construction practices that minimize the environmental impact of the building materials are incorporated, environmentally efficient building systems and materials are incorporated, and waste generated during construction is reduced.

11.601.1 Conditioned floor area. Finished floor area of a dwelling unit after the remodeling is limited. Finished floor area is calculated in accordance with NAHBRC Z765. Only the finished floor area for stories above grade plane is included in the calculation.

		Points
(1)	less than or equal to 1,000 square feet (93 m^2)	**15**
(2)	less than or equal to 1,500 square feet (139 m^2)	**12**
(3)	less than or equal to 2,000 square feet (186 m^2)	**9**
(4)	less than or equal to 2,500 square feet (232 m^2)	**6**
(5)	greater than 4,000 square feet (372 m^2)	**Mandatory**

(For every 100 square feet (9.29 m^2) over 4,000 square feet (372 m^2), one point is to be added the threshold points shown in Table 305.3.7 for each rating level.)

Multi-Unit Building Note: For a multi-unit building, a weighted average of the individual unit sizes is used for this practice.

11.601.2 Material usage. Newly installed structural systems are designed or construction techniques are implemented that reduce and optimize material usage.		**9 Max**
(Points awarded only when the newly installed portion of each structural system comprises at least 25 percent of the total area of that structural system after the remodel)		
(1)	Minimum structural member or element sizes necessary for strength and stiffness in accordance with advanced framing techniques or structural design standards are selected.	3
(2)	Higher-grade or higher-strength of the same materials than commonly specified for structural elements and components in the building are used and element or component sizes are reduced accordingly.	3
(3)	Performance-based structural design is used to optimize lateral force-resisting systems.	3

11.601.3 Building dimensions and layouts. Building dimensions and layouts are designed to reduce material cuts and waste. This practice is used for a minimum of 80 percent of the newly installed areas:

(Points awarded only when the newly installed area of the building comprises at least 25 percent of the total area of that element of the building after the remodel)

(1)	floor area	3
(2)	wall area	3
(3)	roof area	3
(4)	cladding or siding area	3
(5)	penetrations or trim area	1

GREEN BUILDING PRACTICES	POINTS
11.601.4 Framing and structural plans. Detailed framing or structural plans, material quantity lists and on-site cut lists for newly installed framing, structural materials, and sheathing materials are provided.	4

	GREEN BUILDING PRACTICES	POINTS
11.601.5 Prefabricated components. Precut or preassembled components, or panelized or precast assemblies are utilized for a minimum of 90 percent for the following system or building:		13 Max
(Points awarded only when the newly installed system comprises at least 25 percent of the total area of that system of the building after the remodel)		
(1)	floor system	4
(2)	wall system	4
(3)	roof system	4
(4)	modular construction for any new construction located above grade	13

	GREEN BUILDING PRACTICES	POINTS
11.601.6 Stacked stories. Stories above grade are stacked, such as in 1½-story, 2-story, or greater structures. The area of the upper story is a minimum of 50 percent of the area of the story below, based on areas with a minimum ceiling height of 7 feet (2,134 mm).		8 Max
(1)	first stacked story	4
(2)	for each additional stacked story	2

	GREEN BUILDING PRACTICES	POINTS
11.601.7 Site-applied finishing materials. Building materials or assemblies listed below that do not require additional site-applied material for finishing are incorporated in the building.		12 Max
(a)	pigmented, stamped, decorative, or final finish concrete or masonry	
(b)	interior trim not requiring paint or stain	
(c)	exterior trim not requiring paint or stain	

(d)	window, skylight, and door assemblies not requiring paint or stain on one of the following surfaces:
	  i. exterior surfaces
	  ii. interior surfaces	

(e)	interior wall coverings or systems not requiring paint or stain or other type of finishing application	
(f)	exterior wall coverings or systems not requiring paint or stain or other type of finishing application	
(g)	pre-finished hardwood flooring	
(1)	90 percent or more (after the remodel) of the installed building materials or assemblies listed above: **(Points awarded for each type of material or assembly.)**	5
(2)	50 percent to less than 90 percent (after the remodel) of the installed building material or assembly listed above: **(Points awarded for each type of material or assembly.)**	2
(3)	35 percent to less than 50 percent (after the remodel) of the installed building material or assembly listed above: **(Points awarded for each type of material or assembly.)**	1

GREEN BUILDING PRACTICES	POINTS
11.601.8 Foundations. A foundation system that minimizes soil disturbance, excavation quantities and material usage, such as frost-protected shallow foundations, isolated pier and pad foundations, deep foundations, post foundations, or helical piles is selected, designed, and constructed. The foundation is used on 25 percent or more of the building footprint after the remodel.	3

11.602 ENHANCED DURABILITY AND REDUCED MAINTENANCE

11.602.0 Intent. Design and construction practices are implemented that enhance the durability of materials and reduce in-service maintenance.	

11.602.1 Moisture Management – Building Envelope

11.602.1.1 Capillary breaks

11.602.1.1.1 A capillary break and vapor retarder are installed at concrete slabs in accordance with IRC Sections R506.2.2 and R506.2.3 or IBC Sections 1910 and 1805.4.1. **This practice is not mandatory for existing slabs without apparent moisture problem.**	**Mandatory**
11.602.1.1.2 A capillary break to prevent moisture migration into foundation wall is provided between the footing and the foundation wall on all new foundations, and on not less than 25 percent of the total length of the foundation after the remodel.	3
11.602.1.2 Foundation waterproofing. Enhanced foundation waterproofing is installed on all new foundations, and on not less than 25 percent of the total length of the foundation after the remodel using one or both of the following: (1) rubberized coating, or (2) drainage mat	4

11.602.1.3 Foundation drainage.

11.602.1.3.1 Where required by the ICC IRC or IBC for habitable and usable spaces below grade, exterior drain tile is installed. **This practice is not mandatory for existing space without apparent moisture problem.**	**Mandatory**
11.602.1.3.2 Interior and exterior foundation perimeter drains are installed and sloped to discharge to daylight, dry well, or sump pit on all new foundations and not less than 25 percent of the total length of the foundation after the remodel.	4

GREEN BUILDING PRACTICES	POINTS

11.602.1.4 Crawlspaces.

11.602.1.4.1 Vapor retarder for all new unconditioned vented crawlspace foundations and not less than 25 percent of the total area after the remodel is in accordance with the following, as applicable. Joints of vapor retarder overlap a minimum of 6 inches (152 mm) and are taped.

(1)	Floors. Minimum 6 mil vapor retarder installed on the crawlspace floor and extended at least 6 inches up the wall and is attached and sealed to the wall.	6
(2)	Walls. Dampproof walls are provided below finished grade.	Mandatory
	This practice is not mandatory for existing walls without apparent moisture problem.	

11.602.1.4.2 For all new foundations and not less than 25 percent of the total area of the crawlspace after the remodel, crawlspace that is built as a conditioned area is sealed to prevent outside air infiltration and provided with conditioned air at a rate not less than 0.02 cfm (.009 L/s) per square foot of horizontal area and one of the following is implemented:

(1)	a concrete slab over 6 mil polyethylene or polystyrene sheeting, lapped a minimum of 6 inches (152 mm), and taped or sealed at the seams.	8
(2)	6 mil polyethylene sheeting, lapped a minimum of 6 inches (152 mm), and taped at the seams.	Mandatory
	This practice is not mandatory for existing foundations without apparent moisture problem.	

11.602.1.5 Termite barrier. Continuous physical foundation termite barrier used with low toxicity treatment or with no chemical treatment is installed in geographical areas that have subterranean termite infestation potential determined in accordance with Figure 6(3).	4

11.602.1.6 Termite-resistant materials. In areas of termite infestation probability as defined by Figure 6(3), termite-resistant materials are used as follows:

(1)	In areas of slight to moderate termite infestation probability: for the foundation, all structural walls, floors, concealed roof spaces not accessible for inspection, exterior decks, and exterior claddings within the first 2 feet (610 mm) above the top of the foundation.	2
(2)	In areas of moderate to heavy termite infestation probability: for the foundation, all structural walls, floors, concealed roof spaces not accessible for inspection, exterior decks, and exterior claddings within the first 3 feet (914 mm) above the top of the foundation.	4
(3)	In areas of very heavy termite infestation probability: for the foundation, all structural walls, floors, concealed roof spaces not accessible for inspection, exterior decks, and exterior claddings.	6

GREEN BUILDING PRACTICES	POINTS

11.602.1.7 Moisture control measures

11.602.1.7.1 Moisture control measures are in accordance with the following:	
(1) Building materials with visible mold are not installed or are cleaned or encapsulated prior to concealment and closing.	2
(2) Insulation in cavities is dry in accordance with manufacturer's instructions when enclosed (e.g., with drywall).	**Mandatory 2**
(3) The moisture content of lumber is sampled to ensure it does not exceed 19 percent prior to the surface and/or cavity enclosure.	4

11.602.1.7.2 Moisture content of subfloor, substrate, or concrete slabs is in accordance with the appropriate industry standard for the finish flooring to be applied.	2

11.602.1.8 Water-resistive barrier. Where required by the ICC IRC or IBC, a water-resistive barrier and/or drainage plane system is installed behind newly installed exterior veneer and/or siding and where there is evidence of a moisture problem.	**Mandatory**

11.602.1.9 Flashing. Flashing is provided as follows to minimize water entry into wall and roof assemblies and to direct water to exterior surfaces or exterior water-resistive barriers for drainage. Flashing details are provided in the construction documents and are in accordance with the fenestration manufacturer's instructions, the flashing manufacturer's instructions, or as detailed by a registered design professional.

Points awarded only when practices (2)-(7) are implemented in all newly installed construction and not less than 25 percent of the applicable building elements for the entire building after the remodel.

(1) Flashing is installed at all of the following locations, as applicable:	**Mandatory**
(a) around exterior fenestrations, skylights and doors	
(b) at roof valleys	
(c) at all building-to-deck, -balcony, -porch, and -stair intersections	
(d) at roof-to-wall intersections, at roof-to-chimney intersections, at wall-to-chimney intersections, and at parapets.	
(e) at ends of and under masonry, wood, or metal copings and sills	
(f) above projecting wood trim	
(g) at built-in roof gutters, and	
(h) drip edge is installed at eaves and rake edges.	
These practices are not mandatory for existing building elements without apparent moisture problem.	
(2) All window head and jamb flashing is self-adhered flashing complying with AAMA 711-07.	2
(3) Pan flashing is installed at sills of all exterior windows and doors	3
(4) Seamless, preformed kickout flashing, or prefabricated metal with soldered seams is provided at all roof-to-wall intersections. The type and thickness of the material used for roof flashing including but not limited kickout and step flashing is commensurate with the anticipated service life of the roofing material.	3

GREEN BUILDING PRACTICES	POINTS
(5) A rainscreen wall design as follows is used for exterior wall assemblies	**4 Max**
(a) a system designed with minimum ¼-inch air space exterior to the water-resistive barrier, vented to the exterior at top and bottom of the wall and integrated with flashing details, or	**4**
(b) a cladding material or a water-resistive barrier with enhanced drainage, meeting 75 percent drainage efficiency determined in accordance with ASTM E2273.	**2**
(6) Through-wall flashing is installed at transitions between wall cladding materials, or wall construction types.	**2**
(7) Flashing is installed at expansion joints in stucco walls	**2**

11.602.1.10 Exterior doors. Entries at exterior door assemblies, inclusive of side lights, are covered by one of the following methods to protect the building from the effects of precipitation and solar radiation. A projection factor of 0.375 minimum is provided. Eastern- and western-facing entries in Climate Zones 1, 2, and 3, as determined in accordance with Figure 6(1) or Appendix C, have a projection factor of 1.0 minimum, unless protected from direct solar radiation by other means (e.g., screen wall, vegetation).	**2 per exterior door** **6 Max**
(a) installing a porch roof or awning	
(b) extending the roof overhang	
(c) recessing the exterior door	

11.602.1.11 Tile backing materials. Tile backing materials installed under tiled surfaces in wet areas are in accordance with ASTM C1178, C1278, C1288, or C1325. **This practice is not mandatory for existing tile surfaces without apparent moisture problem.**	**Mandatory**

11.602.1.12 Roof overhangs. Roof overhangs, in accordance with Table 11.602.1.12, are provided over a minimum of 90 percent of exterior walls to protect the building envelope.	**4**

Table 11.602.1.12

Minimum Roof Overhang for One- & Two-Story Buildings

Inches of Rainfall [1]	Eave Overhang (Inches)	Rake Overhang (Inches)
≤40	12	12
>41 and ≤70	18	12
>70	24	12

(1) Annual mean total precipitation in inches is in accordance with Figure 6(2).
For SI: 12 inches = 304.8 mm

11.602.1.13 Ice barrier. In areas where there has been a history of ice forming along the eaves causing a backup of water, an ice barrier is installed in accordance with the ICC IRC or IBC at roof eaves of pitched roofs and extends a minimum of 24 inches (610 mm) inside the exterior wall line of the building.	**Mandatory**

GREEN BUILDING PRACTICES	POINTS

11.602.1.14 Architectural features. Architectural features that increase the potential for the water intrusion are avoided:	
(1) All horizontal ledgers are sloped away to provide gravity drainage as appropriate for the application.	**Mandatory** 1
(2) No roof configurations that create horizontal valleys in roof design.	2
(3) No recessed windows and architectural features that trap water on horizontal surfaces.	2

11.602.2 Roof surfaces. A minimum of 90 percent of roof surfaces, not used for roof penetrations and associated equipment, on-site renewable energy systems such as photovoltaics or solar thermal energy collectors, or rooftop decks, amenities and walkways, are constructed of one or both of the following:	3
(1) products that are in accordance with the ENERGY STAR® cool roof certification or equivalent	
(2) a vegetated roof system	

11.602.3 Roof water discharge. A gutter and downspout system or splash blocks and effective grading are provided to carry water a minimum of 5 feet (1524 mm) away from perimeter foundation walls.	4

11.602.4 Finished grade.

11.602.4.1 Finished grade at all sides of a building is sloped to provide a minimum of 6 inches (150 mm) of fall within 10 feet (3048 mm) of the edge of the building. Where lot lines, walls, slopes, or other physical barriers prohibit 6 inches (152 mm) of fall within 10 feet (3048 mm), the final grade is sloped away from the edge of the building at a minimum slope of 2 percent.	**Mandatory**

11.602.4.2 The final grade is sloped away from the edge of the building at a minimum slope of 5 percent.	1

11.602.4.3 Water is directed to drains or swales to ensure drainage away from the structure.	1

11.603
REUSED OR SALVAGED MATERIALS

11.603.0 Intent. Practices that reuse or modify existing structures, salvage materials for other uses, or use salvaged materials in the building's construction are implemented.

11.603.1 Reuse of existing building. Major elements or components of existing buildings and structures are reused, modified, or deconstructed for later use. **(Points awarded for every 200 square feet (18.5 m^2) of floor area.)**	1 **12 Max**

GREEN BUILDING PRACTICES	POINTS
11.603.2 Salvaged materials. Reclaimed and/or salvaged materials and components are used. The total material value and labor cost of salvaged materials is equal to or exceeds 1 percent of the total construction cost. **(Points awarded per 1% of salvaged materials used based on the total construction cost.)** **(Materials, elements, or components awarded points under Section 11.603.1 shall not be awarded points under Section 11.603.2.)**	**1** **9 Max**
11.603.3 Scrap materials. Sorting and reuse of scrap building material is facilitated (e.g., a central storage area or dedicated bins are provided).	4

11.604
RECYCLED-CONTENT BUILDING MATERIALS

GREEN BUILDING PRACTICES	POINTS
11.604.1 Recycled content. Building materials with recycled content are used for two minor and/or two major components of the building.	**Per Table 11.604.1**

Table 11.604.1
Recycled Content

Material Percentage Recycled Content	Points Per 2 Minor	Points Per 2 Major
25% to less than 50%	1	2
50% to less than 75%	2	4
more than 75%	3	6

11.605
RECYCLED CONSTRUCTION WASTE

GREEN BUILDING PRACTICES	POINTS
11.605.0 Intent. Waste generated during construction is recycled. All waste classified as hazardous is properly handled and disposed of.	
11.605.1 Hazardous Waste. The construction waste management plan shall include information on the proper handling and disposal of hazardous waste. All hazardous waste is properly handled and disposed of.	**Mandatory**
11.605.2 Construction waste management plan. A construction waste management plan is developed, posted at the jobsite, and implemented with a goal of recycling or salvaging a minimum of 50 percent (by weight) of construction waste.	6

GREEN BUILDING PRACTICES	POINTS
11.605.3 On-site recycling. On-site recycling measures following applicable regulations and codes are implemented, such as the following:	7
(a) Materials are ground or otherwise safely applied on-site as soil amendment or fill. A minimum of 50 percent (by weight) of construction and land-clearing waste is diverted from landfill.	
(b) Alternative compliance methods approved by the Adopting Entity.	
(c) Compatible untreated biomass material (lumber, posts, beams etc.) are set aside for combustion if a Solid Fuel Burning Appliance per Section 11.901.2.1(2) will be available for on-site renewable energy.	
11.605.4 Recycled construction materials. Construction materials (e.g., wood, cardboard, metals, drywall, plastic, asphalt roofing shingles, or concrete) are recycled offsite.	6 Max
(1) a minimum of two types of materials are recycled	3
(2) for each additional recycled material type	1

11.606 RENEWABLE MATERIALS

11.606.0 Intent. Building materials derived from renewable resources are used.	

		POINTS
11.606.1 Biobased products. The following biobased products are used:		8 Max
	(a) certified solid wood in accordance with Section 11.606.2	
	(b) engineered wood	
	(c) bamboo	
	(d) cotton	
	(e) cork	
	(f) straw	
	(g) natural fiber products made from crops (soy-based, corn-based)	
	(h) products with the minimum biobased contents of the USDA 7 CFR Part 2902	
	(i) other biobased materials with a minimum of 50 percent biobased content (by weight or volume)	
(1)	Two types of biobased materials are used, each for more than 0.5 percent of the project's projected building material cost.	3
(2)	Two types of biobased materials are used, each for more than 1 percent of the project's projected building material cost.	6
(3)	For each additional biobased material used for more than 0.5 percent of the project's projected building material cost.	1 2 Max

GREEN BUILDING PRACTICES	POINTS
11.606.2 Wood-based products. Wood or wood-based products are certified to the requirements of one of the following recognized product programs:	
(a) American Forest Foundation's *American Tree Farm System*® (ATFS)	
(b) Canadian Standards Association's *Sustainable Forest Management System Standards* (CSA Z809)	
(c) *Forest Stewardship Council* (FSC)	
(d) *Program for Endorsement of Forest Certification Systems* (PEFC)	
(e) *Sustainable Forestry Initiative*® *Program (SFI)*	
(f) other product programs mutually recognized by PEFC	
(1) A minimum of two certified wood-based products are used for minor elements of the building (e.g., all trim, cabinetry, or millwork).	3
(2) A minimum of two certified wood-based products are used in major elements of the building (e.g., walls, floors, roof).	4

11.606.3 Manufacturing energy. Materials are used for major components of the building that are manufactured using a minimum of 33 percent of the primary manufacturing process energy derived from renewable sources, combustible waste sources, or renewable energy credits (RECs). **(2 points awarded per material.)**	6 Max

11.607
RECYCLING AND WASTE REDUCTION

11.607.1 Recycling. Recycling by the occupant is facilitated by one or more of the following methods:	
(1) A built-in collection space in each kitchen and an aggregation/pick-up space in a garage, covered outdoor space, or other area for recycling containers is provided.	3
(2) Compost facility is provided on the site.	3

11.607.2 Food waste disposers. A minimum of one food waste disposer is installed at the primary kitchen sink.	1

11.608
RESOURCE-EFFICIENT MATERIALS

11.608.1 Resource-efficient materials. Products containing fewer materials are used to achieve the same end-use requirements as conventional products, including but not limited to:	9 Max 3 per each material
(1) lighter, thinner brick with bed depth less than 3 inches and/or brick with coring of more that 25 percent	
(2) engineered wood or engineered steel products	
(3) roof or floor trusses	

GREEN BUILDING PRACTICES	POINTS

11.609 REGIONAL MATERIALS

11.609.1 Regional materials. Regional materials are used for major elements or components of the building.	**10 Max 2 per each material type**

11.610 LIFE CYCLE ANALYSIS

11.610.1 Life cycle analysis. A life cycle analysis (LCA) tool is used to select environmentally preferable products or assemblies, or an LCA is conducted on the entire building. Points are awarded in accordance with Section 11.610.1.1 or 11.610.1.2. Only one method of analysis or tool may be utilized. A reference service life for the building is 60 years for any life cycle analysis tool. Results of the LCA are reported in the manual required in Section 11.1003.1(1) of this Standard in terms of the environmental impacts listed in this practice and it states if operating energy was included in the LCA.	**15 Max**
11.610.1.1 Whole-building life cycle analysis. A whole-building LCA is performed using a life cycle assessment and data compliant with ISO 14044 or other recognized standards.	**15**
11.610.1.2 Life cycle analysis for a product or assembly. An environmentally preferable product or assembly is selected for an application based upon the use of an LCA tool that incorporates data methods compliant with ISO 14044 or other recognized standards that compare the environmental impact of products or assemblies.	**10 Max**
11.610.1.2.1 Product LCA. A product with improved environmental impact measures compared to another product(s) intended for the same use is selected. The environmental impact measures used in the assessment are selected from the following:	**Per Table 11.610.1.2.1 10 Max**

    (a)   Fossil fuel consumption

    (b)   Global warming potential

    (c)   Acidification potential

    (d)   Eutrophication potential

    (e)   Ozone depletion potential

(Points awarded for each product/system comparison where the selected product/system improved upon the environmental impact measures by an average of 15 percent.)

Table 11.610.1.2.1
Product LCA

4 Impact Measures	5 Impact Measures
POINTS	
2	3

GREEN BUILDING PRACTICES	POINTS

11.610.1.2.2 Assembly LCA. An assembly with improved environmental impact measures compared to a functionally comparable assembly is selected. The full life cycle, from resource extraction to demolition and disposal (including but not limited to on-site construction, maintenance and replacement, material and product embodied acquisition, and process and transportation energy), is assessed. The assessment does not include electrical and mechanical equipment and controls, plumbing products, fire detection and alarm systems, elevators, and conveying systems. The following functional building elements are eligible for points under this practice:	**Per Table 11.610.1.2.2 10 Max**

(a)   exterior walls

(b)   roof/ceiling

(c)   interior walls or ceilings

(d)   intermediate floors

The environmental impact measures used in the assessment are selected from the following:

(a)   Fossil fuel consumption

(b)   Global warming potential

(c)   Acidification potential

(d)   Eutrophication potential

(e)   Ozone depletion potential

(Points are awarded based on the number of functional building elements that improve upon environmental impact measures by an average of 15 percent.)

Table 11.610.1.2.2
Assembly LCA

	4 Impact Measures	5 Impact Measures
	POINTS	
2 functional building elements	3	6
3 functional building elements	4	8
4 functional building elements	5	10

11.611
INNOVATIVE PRACTICES

11.611.1 Manufacturer's environmental management system concepts. Product manufacturer's operations and business practices include environmental management system concepts, and the production facility is registered to ISO 14001 or equivalent. The aggregate value of building products from registered ISO 14001 or equivalent production facilities is 1 percent or more of the estimated total building materials cost. **(1 point awarded per percent.)**	**10 Max**

GREEN BUILDING PRACTICES	POINTS
11.611.2 Sustainable products. One or more of the following products are used for at least 30% of the floor or wall area of the entire dwelling unit, as applicable. Products are certified by a third-party agency accredited to ISO Guide 65.	9 Max
(1) 50% or more of carpet installed (by square feet) is certified to NSF 140.	3
(2) 50% or more of resilient flooring installed (by square feet) is certified to NSF 332.	3
(3) 50% or more of the insulation installed (by square feet) is certified to EcoLogo CCD-016.	3
(4) 50% or more of interior wall coverings installed (by square feet) is certified to NSF 342	3
(5) 50% or more of the gypsum board installed (by square feet) is certified to ULE ISR 100.	3
(6) 50% or more of the door leafs installed (by number of door leafs) is certified to ULE ISR 102.	3
(7) 50% or more of the tile installed (by square feet) is certified to TCNA A138.1 Specifications for Sustainable Ceramic Tiles, Glass Tiles and Tile Installation Materials.	3

GREEN BUILDING PRACTICES	POINTS
11.611.3 Universal design elements. Dwelling incorporates one or more of the following universal design elements. Conventional industry tolerances are permitted.	10 Max
(1) Any no-step entrance into the dwelling which (1) is accessible from a substantially level parking or drop-off area (no more than 2%) via an accessible path which has no individual change in elevation or other obstruction of more than 1-1/2 inches in height with the pitch not exceeding 1 in 12 and (2) provides a minimum 32-inch wide clearance into the dwelling.	3
(2) Minimum 36-inch wide accessible route from the no-step entrance into at least one visiting room in the dwelling and into at least one full or half bathroom which has a minimum 32 inch clear door width and a 30-inch by 48-inch clear area inside the bathroom outside the door swing.	3
(3) Minimum 36-inch-wide accessible route from the no-step entrance into at least one bedroom which has a minimum 32-inch clear door width.	3
(4) Blocking or equivalent installed in the accessible bathroom walls for future installation of grab bars at water closet and bathing fixture, if applicable.	1

11.701
MINIMUM ENERGY EFFICIENCY REQUIREMENTS

11.701.4 Mandatory practices.

11.701.4.1 HVAC systems.

11.701.4.1.1 HVAC system sizing. Newly installed or modified space heating and cooling system is sized according to heating and cooling loads calculated using ACCA Manual J, or equivalent. New equipment is selected using ACCA Manual S or equivalent.	Mandatory

11.701.4.1.2 Radiant and hydronic space heating. Where installed as a primary heat source in the building, new radiant or hydronic space heating system is designed using industry-approved guidelines and standards (e.g., ACCA Manual J, AHRI I=B=R, ANSI/ACCA 5 QI-2010, or an accredited design professional's and manufacturer's recommendations).	Mandatory

GREEN BUILDING PRACTICES	POINTS

11.701.4.2 Duct systems.

11.701.4.2.1 Duct air sealing. Newly installed, modified, or ducts that are exposed during the remodel are air sealed. All duct sealing materials are in conformance with UL 181A or UL 181B specifications and are installed in accordance with manufacturer's instructions.	**Mandatory**
11.701.4.2.2 Supply ducts. Building cavities are not used as supply ducts. Existing building cavities currently used as supply ducts exposed during the remodel are lined.	**Mandatory**
11.701.4.2.3 Duct system sizing. New or modified Duct system is sized and designed in accordance with ACCA Manual D or equivalent.	**Mandatory**

11.701.4.3 Insulation and air sealing.

11.701.4.3.1 Building Thermal Envelope. The building thermal envelope exposed or created during the remodel is durably sealed to limit infiltration. The sealing methods between dissimilar materials allow for differential expansion and contraction. The following are caulked, gasketed, weather-stripped or otherwise sealed with an air barrier material, suitable film or solid material:	**Mandatory**

(a)	All joints, seams and penetrations.	
(b)	Site-built windows, doors and skylights.	
(c)	Openings between window and door assemblies and their respective jambs and framing.	
(d)	Utility penetrations.	
(e)	Dropped ceilings or chases adjacent to the thermal envelope.	
(f)	Knee walls.	
(g)	Walls and ceilings separating a garage from conditioned spaces.	
(h)	Behind tubs and showers on exterior walls.	
(i)	Common walls between dwelling units.	
(j)	Attic access openings.	
(k)	Rim joist junction.	
(l)	Other sources of infiltration.	

11.701.4.3.2 Air sealing and insulation. Grade 3 insulation installation is not permitted. The compliance of the building envelope air tightness and insulation installation is demonstrated in accordance with Section 11.701.4.3.2(1) or 11.701.4.3.2(2).	**Mandatory**

(1) **Testing option.** Building envelope tightness and insulation installation is considered acceptable when air leakage is less than seven air changes per hour (ACH) when tested with a blower door at a pressure of 33.5 psf (50 Pa). Testing is conducted after rough-in and after installation of penetrations of the building envelope, including penetrations for utilities, plumbing, electrical, ventilation and combustion appliances. Testing is conducted under the following conditions:

    (a) Exterior windows and doors, fireplace and stove doors are closed, but not sealed;

GREEN BUILDING PRACTICES	POINTS

(b) Dampers are closed, but not sealed, including exhaust, intake, makeup air, backdraft, and flue dampers;

(c) Interior doors are open;

(d) Exterior openings for continuous ventilation systems and heat recovery ventilators are closed and sealed;

(e) Heating and cooling system(s) is turned off;

(f) HVAC ducts terminations are not sealed; and

(g) Supply and return registers are not sealed.

(2) **Visual inspection option.** Building envelope tightness and insulation installation are considered acceptable when the items listed in Table 11.701.4.3.2(2) applicable to the method of construction and exposed and visible during the remodel, are field verified.

Table 11.701.4.3.2(2)
Air Barrier and Insulation Inspection Component Criteria

COMPONENT	CRITERIA
Air barrier and thermal barrier	• Exterior thermal envelope insulation for framed walls is installed in substantial contact and continuous alignment with building envelope air barrier. • Breaks or joints in the air barrier are filled or repaired. • Air-permeable insulation is not used as a sealing material. • Air-permeable insulation is installed with an air barrier.
Ceiling/attic	• Air barrier in dropped ceiling/soffit is substantially aligned with insulation and any gaps are sealed. • Attic access (except unvented attic), knee wall door, or drop-down stair is sealed.
Exterior walls	• Corners and headers are insulated. • Junction of foundation and sill plate is sealed.
Windows and doors	• Space between window/door jambs and framing is sealed.
Rim joists	• Rim joists are insulated and include an air barrier.
Floors (including above-garage and cantilevered floors)	• Insulation is installed to maintain permanent contact with underside of subfloor decking. • Air barrier is installed at any exposed edge of insulation.
Crawlspace walls	• Where installed, insulation is permanently attached to walls. • Exposed earth in unvented crawlspaces is covered with Class I vapor retarder with overlapping joints taped.
Shafts, penetrations	• Duct shafts, flue shafts, and utility penetrations, opening to the exterior or an unconditioned space are sealed.
Narrow cavities	• Batts in narrow cavities are cut to fit, or narrow cavities are filled by sprayed/blown insulation.
Garage separation	• Air sealing is provided between the garage and conditioned spaces.
Recessed lighting	• Recessed light fixtures not installed in the conditioned space are air tight, IC rated, and sealed to drywall.
Plumbing and wiring	• Insulation is placed between the outside and pipes. Batt insulation is cut to fit around wiring and plumbing, or sprayed/blown insulation extends behind piping and wiring.
Shower/tub adjacent to exterior wall	• Showers and tubs adjacent to exterior walls have insulation and an air barrier separation from the exterior.
Electrical/phone box in exterior walls	• Air barrier extends behind boxes or air sealed-type boxes are installed.
Common wall	• Air barrier is installed in common walls between dwelling units.
HVAC register boots	• HVAC register boots that penetrate building envelope are sealed to subfloor or drywall.
Fireplace	• Fireplace walls include an air barrier.

GREEN BUILDING PRACTICES	POINTS
11.701.4.3.3 Fenestration air leakage. Newly installed Windows, skylights and sliding glass doors have an air infiltration rate of no more than 0.3 cfm per square foot (1.5 L/s/m^2), and swinging doors no more than 0.5 cfm per square foot (2.6 L/s/m^2), when tested in accordance with NFRC 400 or AAMA/WDMA/CSA 101/I.S.2/A440 by an accredited, independent laboratory and listed and labeled. This practice does not apply to site-built windows, skylights, and doors.	**Mandatory**
11.701.4.3.4 Recessed lighting. Newly installed recessed luminaires installed in the building thermal envelope are sealed to limit air leakage between conditioned and unconditioned spaces. All recessed luminaires are IC-rated and labeled as meeting ASTM E 283 when tested at 1.57 psf (75 Pa) pressure differential with no more than 2.0 cfm (0.944 L/s) of air movement from the conditioned space to the ceiling cavity. All recessed luminaires are sealed with a gasket or caulk between the housing and the interior of the wall or ceiling covering.	**Mandatory**
11.701.4.4 High-efficacy lighting. A minimum of 50 percent of the newly installed hard-wired lighting fixtures, or the bulbs in those fixtures, qualify as high efficacy or equivalent.	**Mandatory**
11.701.4.5 Boiler supply piping. Boiler supply piping in unconditioned space that is accessible during the remodel is insulated.	**Mandatory**

11.901
POLLUTANT SOURCE CONTROL

11.901.0 Intent. Pollutant sources are controlled.

11.901.1 Space and water heating options

11.901.1.1 Natural draft furnaces, boilers, or water heaters are not located in conditioned spaces, including conditioned crawlspaces, unless located in a mechanical room that has an outdoor air source, and is sealed and insulated to separate it from the conditioned space(s). **(Points are awarded only for buildings that use natural draft combustion space or water heating equipment.)**	5
11.901.1.2 Air handling equipment or return ducts are not located in the garage, unless placed in isolated, air-sealed mechanical rooms with an outside air source.	5

11.901.1.3 The following combustion space heating or water heating equipment is installed within conditioned space:

(1)	all furnaces or all boilers		
	(a)	power vent furnace(s) or boiler(s)	3
	(b)	direct vent furnace(s) or boiler(s)	5
(2)	all water heaters		
	(a)	power vent water heater(s)	3
	(b)	direct vent water heater(s)	5

GREEN BUILDING PRACTICES	POINTS
11.901.1.4 Newly installed gas-fired fireplaces and direct heating equipment is listed and is installed in accordance with the NFPA 54,ICC IFG, or the applicable local gas appliance installation code. Gas-fired fireplaces and direct heating equipment are vented to the outdoors.	**Mandatory**

11.901.1.5 Natural gas and propane fireplaces are direct vented have permanently fixed glass fronts or gasketed doors, and comply with CSA Z21.88/CSA 2.33 or CSA Z21.50/CSA 2.22.	7

11.901.1.6 The following electric equipment is installed:	
(1) heat pump air handler in unconditioned space	**2**
(2) heat pump air handler in conditioned space	**5**

11.901.2 Solid fuel-burning appliances.

11.901.2.1 Newly installed solid fuel-burning fireplaces, inserts, stoves and heaters are code compliant and are in accordance with the following requirements:	**Mandatory**
(1) Site-built masonry wood-burning fireplaces are equipped with outside combustion air and a means of sealing the flue and the combustion air outlets to minimize interior air (heat) loss when not in operation.	
(2) Factory-built, wood-burning fireplaces are in accordance with the certification requirements of UL 127 and are EPA certified.	
(3) Wood stove and fireplace inserts, as defined in UL 1482 Section 3.8, are in accordance with the certification requirements of UL 1482 and are in accordance with the emission requirements of the EPA Certification and the State of Washington WAC 173-433-100(3).	
(4) Pellet (biomass) stoves and furnaces are in accordance with the requirements of ASTM E1509 or are EPA certified.	
(5) Masonry heaters are in accordance with the definitions in ASTM E1602 and ICC IBC, Section 2112.1.	
(6) Removal of or rendering unusable an existing fireplace or fuel burning appliance that is not in accordance with 11.901.2.1 or replacement of each fireplace or appliance that is not in accordance with 11.901.2.1 with a compliant appliance.	

11.901.2.2 Fireplaces, woodstoves, pellet stoves, or masonry heaters are not installed.	7

GREEN BUILDING PRACTICES	POINTS

11.901.3 Garages. Garages are in accordance with the following:	
(1) Attached garage	
(a) Where installed in the common wall between the attached garage and conditioned space, the door is tightly sealed and gasketed.	**Mandatory 2**
(b) A continuous air barrier is provided between walls and ceilings separating the garage space from the conditioned living spaces.	**Mandatory 2**
(c) For one- and two-family dwelling units, a 100 cfm (47 L/s) or greater ducted, or 70 cfm (33 L/s) cfm or greater unducted wall exhaust fan is installed and vented to the outdoors, designed and installed for continuous operation, or has controls (e.g., motion detectors, pressure switches) that activate operation for a minimum of 1 hour when either human passage door or roll-up automatic doors are operated. For ducted exhaust fans, the fan airflow rating and duct sizing are in accordance with Appendix A.	8
(2) A carport is installed, the garage is detached from the building, or no garage is installed.	10

11.901.4 Wood materials. A minimum of 85 percent of newly installed material within a product group (i.e., wood structural panels, countertops, composite trim/doors, custom woodwork, and/or component closet shelving) is manufactured in accordance with the following:	**10 Max**
(1) Structural plywood used for floor, wall, and/or roof sheathing is compliant with DOC PS 1 and/or DOC PS 2. OSB used for floor, wall, and/or roof sheathing is compliant with DOC PS 2. The panels are made with moisture-resistant adhesives. The trademark indicates these adhesives as follows: Exposure 1 or Exterior for plywood, and Exposure 1 for OSB.	**Mandatory**
(2) Particleboard and MDF (medium density fiberboard) is manufactured and labeled in accordance with CPA A208.1 and CPA A208.2, respectively. **(Points awarded per product group.)**	2
(3) Hardwood plywood in accordance with HPVA HP-1. **(Points awarded per product group.)**	2
(4) Particleboard, MDF, or hardwood plywood is in accordance with CPA 3. **(Points awarded per product group.)**	3
(5) Composite wood or agrifiber panel products contain no added urea-formaldehyde or are in accordance with the CARB *Composite Wood Air Toxic Contaminant Measure Standard*. **(Points awarded per product group.)**	4
(6) Non-emitting products. **(Points awarded per product group.)**	4

11.901.5 Cabinets. A minimum of 85 percent of newly installed cabinets are in accordance with one or both of the following: **(Where both of the following practices are used, only three points are awarded.)**	
(1) All parts of the cabinet are made of solid wood or non-formaldehyde emitting materials such as metal or glass.	5
(2) The composite wood used in wood cabinets are in accordance with CARB Composite Wood Air Toxic Contaminant Measure Standard or equivalent as certified by a program such as but not limited to, those in Appendix D.	3

ICC 700-2012 NATIONAL GREEN BUILDING STANDARD™

GREEN BUILDING PRACTICES	POINTS

11.901.6 Carpets. Carpets are in accordance with the following:

(1)	Wall-to-wall carpeting is not installed adjacent to water closets and bathing fixtures.	**Mandatory**
(2)	Newly installed carpet area is at least 10 percent of the conditioned floor space and a minimum of 85 percent of newly installed carpet area and carpet cushion (padding) are in accordance with the emission levels of CDPH/EHLB Standard Method v1.1, footnote b in Table 4.1 does not apply (i.e., maximum allowable formaldehyde concentration is 16.5 µg/m^3 (13.5 ppb)). Emission levels are determined by a laboratory accredited to ISO/IEC 17025 and the CDPH/EHLB Standard Method v1.1 is in its scope of accreditation. The product is certified by a third-party program accredited to ISO Guide 65, such as, but not limited to, those in Appendix D.	
	(a) carpet	6
	(b) carpet cushion	2

11.901.7 Hard-surface flooring. A minimum of 10 percent of the conditioned floor space has pre-finished hard-surface flooring installed and at least 85 percent of all prefinished installed hard-surface flooring is in accordance with the emission concentration limits of CDPH/EHLB Standard Method v1.1, footnote b in Table 4.1 does not apply (i.e., maximum allowable formaldehyde concentration is 16.5 µg/m^3 (13.5 ppb)). Emission levels are determined by a laboratory accredited to ISO/IEC 17025 and the CDPH/EHLB Standard Method v1.1 is in its scope of accreditation. The product is certified by a third-party program accredited to ISO Guide 65, such as, but not limited to, those found in Appendix D. Where post-manufacture coatings or surface applications have not been applied, the following hard-surface flooring types are deemed to comply with the emission requirements of this section:	6
(a) Ceramic tile flooring	
(b) Organic-free, mineral-based flooring	
(c) Clay masonry flooring	
(d) Concrete masonry flooring	
(e) Concrete flooring	
(f) Metal flooring	
(g) Glass	

11.901.8 Wall coverings. When at least 10 percent of the interior wall surfaces are covered, a minimum of 85 percent of wall coverings are in accordance with the emission concentration limits of CDPH/EHLB Standard Method v1.1, footnote b in Table 4.1 does not apply (i.e., maximum allowable formaldehyde concentration is 16.5 µg/m^3 (13.5 ppb)). Emission levels are determined by a laboratory accredited to ISO/IEC 17025 and the CDPH/EHLB Standard Method v1.1 is in its scope of accreditation. The product is certified by a third-party program accredited to ISO Guide 65, such as, but not limited to, those in Appendix D.	4

11.901.9 Interior architectural coatings. A minimum of 85 percent of newly applied interior architectural coatings are in accordance with either Section 11.901.9.1 or Section 11.901.9.3, not both. A minimum of 85 percent of architectural colorants are in accordance with Section 11.901.9.2.

GREEN BUILDING PRACTICES	POINTS
11.901.9.1 Site-applied interior architectural coatings, which are inside the waterproofing envelope, are in accordance with one or more of the following:	5

(1) Zero VOC as determined by EPA Method 24 (VOC content below the detection limit for the method)

(2) GreenSeal GS-11

(3) CARB *Suggested Control Measure for Architectural Coatings* (see Table 11.901.9.1)

Table 11.901.9.1
VOC Content Limits For Architectural Coatings[a,b,c]

Coating Category	LIMIT[d] (g/l)
Flat Coatings	50
Non-flat Coatings	100
Non-flat - High Gloss Coatings	150
Specialty Coatings:	
Aluminum Roof Coatings	400
Basement Specialty Coatings	400
Bituminous Roof Coatings	50
Bituminous Roof Primers	350
Bond Breakers	350
Concrete Curing Compounds	350
Concrete/Masonry Sealers	100
Driveway Sealers	50
Dry Fog Coatings	150
Faux Finishing Coatings	350
Fire Resistive Coatings	350
Floor Coatings	100
Form-Release Compounds	250
Graphic Arts Coatings (Sign Paints)	500
High Temperature Coatings	420
Industrial Maintenance Coatings	250
Low Solids Coatings	120[e]
Magnesite Cement Coatings	450
Mastic Texture Coatings	100
Metallic Pigmented Coatings	500
Multi-Color Coatings	250
Pre-Treatment Wash Primers	420
Primers, Sealers, and Undercoaters	100
Reactive Penetrating Sealers	350
Recycled Coatings	250
Roof Coatings	50
Rust Preventative Coatings	250
Shellacs, Clear	730
Shellacs, Opaque	550
Specialty Primers, Sealers, and Undercoaters	100
Stains	250
Stone Consolidants	450
Swimming Pool Coatings	340

GREEN BUILDING PRACTICES	POINTS

Coating Category	LIMIT[d] (g/l)
Traffic Marking Coatings	100
Tub and Tile Refinish Coatings	420
Waterproofing Membranes	250
Wood Coatings	275
Wood Preservatives	350
Zinc-Rich Primers	340

a. The specified limits remain in effect unless revised limits are listed in subsequent columns in the table.
b. Values in this table are derived from those specified by the California Air Resources Board, Architectural Coatings Suggested Control Measure, February 1, 2008.
c. Table 11.901.9.1 architectural coating regulatory category and VOC content compliance determination shall conform to the California Air Resources Board Suggested Control Measure for Architectural Coatings dated February 1, 2008.
d. Limits are expressed as VOC Regulatory (except as noted), thinned to the manufacturer's maximum thinning recommendation, excluding any colorant added to tint bases.
e. Limit is expressed as VOC actual.

11.901.9.2 Architectural coating colorant additive VOC content is in accordance with Table 11.901.9.2. **(Points for 11.901.9.2 are awarded only if base architectural coating is in accordance with 11.901.9.1.)**	1

Table 11.901.9.2
VOC content limits for colorants

Colorant	LIMIT (g/l)
Architectural Coatings, excluding IM Coatings	50
Solvent-Based IM	600
Waterborne IM	50

11.901.9.3 Site-applied interior architectural coatings are in accordance with the emission levels of CDPH/EHLB Standard Method v1.1, footnote b in Table 4.1 does not apply (i.e., maximum allowable formaldehyde concentration is 16.5 µg/m^3 (13.5 ppb)). Emission levels are determined by a laboratory accredited to ISO/IEC 17025 and the CDPH/EHLB Standard Method v1.1 is in its scope of accreditation. The product is certified by a third-party program accredited to ISO Guide 65, such as, but not limited to, those found in Appendix D.	8

11.901.9.4 When the building is occupied during the remodel, a minimum of 85 percent of the newly applied interior architectural coatings are in accordance with either 11.901.9.1 or 11.901.9.3.	Mandatory

GREEN BUILDING PRACTICES	POINTS

11.901.10 Adhesives and sealants. Interior low-VOC adhesives and sealants located inside the water proofing envelope: A minimum of 85 percent of newly applied site-applied products used within the interior of the building are in accordance with one of the following, as applicable.		
(1)	The emission levels of CDPH/EHLB Standard Method v1.1, footnote b in Table 4.1 does not apply (i.e., maximum allowable formaldehyde concentration is 16.5 µg/m³ (13.5 ppb)). Emission levels are determined when tested by a laboratory accredited to ISO/IEC 17025 and the CDPH/EHLB Standard Method v1.1 is in its scope of accreditation. The product is certified by a third-party program accredited to ISO Guide 65, such as, but not limited to, those found in Appendix D.	8
(2)	GreenSeal GS-36	5
	OR	
(3)	SCAQMD Rule 1168 in accordance with Table 11.901.10(3), excluding products that are sold in 16 ounce containers or less and are regulated by the California Air Resources Board (CARB) Consumer Products Regulation.	5

Table 11.901.10(3)
Site Applied Adhesive And Sealants VOC Limits[a,b]

ADHESIVE OR SEALANT	VOC LIMIT (g/l)
Indoor carpet adhesives	50
Carpet pad adhesives	50
Outdoor carpet adhesives	150
Wood flooring adhesive	100
Rubber floor adhesives	60
Subfloor adhesives	50
Ceramic tile adhesives	65
VCT and asphalt tile adhesives	50
Drywall and panel adhesives	50
Cove base adhesives	50
Multipurpose construction adhesives	70
Structural glazing adhesives	100
Single ply roof membrane adhesives	250
Architectural sealants	250

Architectural sealant primer
  Non-porous
  Porous
	250
	775

Modified bituminous sealant primer	500
Other sealant primers	750
CPVC solvent cement	490
PVC solvent cement	510
ABS solvent cement	325
Plastic cement welding	250
Adhesive primer for plastic	550
Contact adhesive	80
Special purpose contact adhesive	250
Structural wood member adhesive	140

a. VOC limit less water and less exempt compounds in grams/liter
b. For low-solid adhesives and sealants, the VOC limit is expressed in grams/liter of material as specified in Rule 1168. For all other adhesives and sealants, the VOC limits are expressed as grams of VOC per liter of adhesive or sealant less water and less exempt compounds as specified in Rule 1168.

GREEN BUILDING PRACTICES	POINTS
11.901.11 Insulation. Emissions of 85 percent of newly installed wall, ceiling, and floor insulation materials are in accordance with the emission levels of CDPH/EHLB. Standard Method v1.1, footnote b in Table 4.1 does not apply (i.e., maximum allowable formaldehyde concentration is 16.5 µg/m^3 (13.5 ppb)). Emission levels are determined by a laboratory accredited to ISO/IEC 17025 and the CDPH/EHLB Standard Method v1.1 is in its scope of accreditation. The product is certified by a third-party program accredited to ISO Guide 65, such as, but not limited to, those in Appendix D.	4
11.901.12 Carbon monoxide (CO) alarms. Where not required by local codes, a carbon monoxide (CO) alarm is installed in a central location outside of each separate sleeping area in the immediate vicinity of the bedrooms. The CO alarm(s) is located in accordance with NFPA 720 and is hard-wired with a battery back-up. The alarm device(s) is certified by a third-party for conformance to either CSA 6.19 or UL 2034.	3
11.901.13 Building entrance pollutants control. Pollutants are controlled at all main building entrances by one of the following methods:	
(1) Exterior grilles or mats are installed in a fixed manner and may be removable for cleaning.	1
(2) Interior grilles or mats are installed in a fixed manner and may be removable for cleaning.	1
11.901.14 Non-smoking areas. Environmental tobacco smoke is minimized by one or more of the following:	
(1) All interior common areas of a multi-unit building are designated as non-smoking areas with posted signage.	1
(2) Exterior smoking areas of a multi-unit building are designated with posted signage and located a minimum of 25 feet from entries, outdoor air intakes, and operable windows.	1
11.901.15 Lead-safe work practices. For buildings constructed before 1978, lead-safe work practices are used during the remodeling.	Mandatory

11.902
POLLUTANT CONTROL

11.902.0 Intent. Pollutants generated in the building are controlled.	

11.902.1 Spot ventilation.	

11.902.1.1 Spot ventilation is in accordance with the following:	
(1) Bathrooms are vented to the outdoors. The minimum ventilation rate is 50 cfm (23.6 L/s) for intermittent operation or 20 cfm (9.4 L/s) for continuous operation in bathrooms. **(Points are awarded only if a window complying with IRC Section R303.3 is provided in addition to mechanical ventilation.)**	Mandatory 1
(2) Clothes dryers are vented to the outdoors.	Mandatory
(3) Kitchen exhaust units and/or range hoods are ducted to the outdoors and have a minimum ventilation rate of 100 cfm (47.2 L/s) for intermittent operation or 25 cfm (11.8 l/s) for continuous operation.	8

GREEN BUILDING PRACTICES	POINTS
11.902.1.2 Bathroom and/or laundry exhaust fan is provided with an automatic timer and/or humidistat:	**11 Max**
(1) for first device	5
(2) for each additional device	2

11.902.1.3 Kitchen range, bathroom, and laundry exhaust are verified to air flow specification. Ventilation airflow at the point of exhaust is tested to a minimum of:	8
(a) 100 cfm (47.2 L/s) intermittent or 25 cfm (11.8 L/s) continuous for kitchens, and	6
(b) 50 cfm (23.6 L/s) intermittent or 20 cfm (9.4 L/s) continuous for bathrooms and/or laundry.	

11.902.1.4 Exhaust fans are ENERGY STAR, as applicable.	**12 Max**
(1) ENERGY STAR, or equivalent, fans **(Points awarded per fan.)**	2
(2) ENERGY STAR, or equivalent, fans operating at or below 1 sone **(Points awarded per fan.)**	3

11.902.2 Building ventilation systems

11.902.2.1 One of the following whole building ventilation systems is implemented and is in accordance with the specifications of Appendix B.	**Mandatory where the maximum air infiltration rate is less than 5 ACH50**
(1) exhaust or supply fan(s) ready for continuous operation and with appropriately labeled controls	3
(2) balanced exhaust and supply fans with supply intakes located in accordance with the manufacturer's guidelines so as to not introduce polluted air back into the building	6
(3) heat-recovery ventilator	7
(4) energy-recovery ventilator	8

11.902.2.2 Ventilation airflow is tested to achieve the design fan airflow at point of exhaust in accordance with Section 11.902.2.1.	4

11.902.2.3 MERV filters 8 or greater are installed on central forced air systems and are accessible. Designer or installer is to verify that the HVAC equipment is able to accommodate the greater pressure drop of MERV 8 filters.	3

GREEN BUILDING PRACTICES	POINTS
11.902.3 Radon control. Radon control measures are in accordance with ICC IRC Appendix F. Zones are defined in Figure 9(1). This practice is not mandatory if the existing building has been tested for radon and is accordance with federal and local acceptable limits.	
(1) Buildings located in Zone 1	**Mandatory**
(a) a passive radon system is installed	7
(b) an active radon system is installed	10
(2) Buildings located in Zone 2 or Zone 3	
(a) a passive or active radon system is installed	7
11.902.4 HVAC system protection. One of the following HVAC system protection measures is performed.	3
(1) HVAC supply registers (boots), return grilles, and rough-ins are covered during construction activities to prevent dust and other pollutants from entering the system.	
(2) Prior to owner occupancy, HVAC supply registers (boots), return grilles, and duct terminations are inspected and vacuumed. In addition, the coils are inspected and cleaned and the filter is replaced if necessary.	
11.902.5 Central vacuum systems. Central vacuum system is installed and vented to the outside.	3
11.902.6 Living space contaminants. The living space is sealed in accordance with Section 11.701.4.3.1 to prevent unwanted contaminants.	**Mandatory**

11.903
MOISTURE MANAGEMENT: VAPOR, RAINWATER, PLUMBING, HVAC

11.903.0 Intent. Moisture and moisture effects are controlled.	

11.903.1 Plumbing	

11.903.1.1 Cold water pipes in unconditioned spaces are insulated to a minimum of R-4 with pipe insulation or other covering that adequately prevents condensation.	2

11.903.1.2 Plumbing is not installed in unconditioned spaces.	5

11.903.2 Duct insulation. Ducts are in accordance with one of the following:	
(1) All HVAC ducts, plenums, and trunks in are conditioned space.	1
(2) All HVAC ducts, plenums, and trunks in are conditioned space. All HVAC ducts are insulated to a minimum of R4.	3

GREEN BUILDING PRACTICES	POINTS
11.903.3 Relative humidity. In climate zones 1A, 2A, 3A, 4A, and 5A as defined by Figure 6(1), equipment is installed to maintain relative humidity (RH) at or below 60 percent using one of the following: **(Points not awarded in other climate zones.)**	7
(1) additional dehumidification system(s)	
(2) central HVAC system equipped with additional controls to operate in dehumidification mode	

11.904
INNOVATIVE PRACTICES

	POINTS
11.904.1 Humidity monitoring system. A humidity monitoring system is installed with a mobile base unit that displays readings of temperature and relative humidity. The system has a minimum of two remote sensor units. One remote unit is placed permanently inside the conditioned space in a central location, excluding attachment to exterior walls, and another remote sensor unit is placed permanently outside of the conditioned space.	2
11.904.2 Kitchen exhaust. A kitchen exhaust unit(s) that equals or exceeds 400 cfm (189 L/s) is installed, and makeup air is provided.	2

11.1001
BUILDING OWNERS' MANUAL FOR ONE- AND TWO-FAMILY DWELLINGS

	POINTS
11.1001.0 Intent. Information on the building's use, maintenance, and green components is provided.	
11.1001.1 A building owner's manual is provided that includes the following, as available and applicable. **(Points awarded per two items. Points awarded for both mandatory and non-mandatory items.)**	1 8 Max
(1) A green building program certificate or completion document.	Mandatory
(2) List of green building features (can include the national green building checklist).	Mandatory
(3) Product manufacturer's manuals or product data sheet for newly installed major equipment, fixtures, and appliances. If product data sheet is in the building owners' manual, manufacturer's manual may be attached to the appliance in lieu of inclusion in the building owners' manual.	Mandatory
(4) Maintenance checklist.	
(5) Information on local recycling programs.	
(6) Information on available local utility programs that purchase a portion of energy from renewable energy providers.	
(7) Explanation of the benefits of using energy-efficient lighting systems [e.g., compact fluorescent light bulbs, light emitting diode (LED)] in high-usage areas.	
(8) A list of practices to conserve water and energy.	

GREEN BUILDING PRACTICES	POINTS

(9)	Local public transportation options.	
(10)	A diagram showing the location of safety valves and controls for major building systems.	
(11)	Where frost-protected shallow foundations are used, owner is informed of precautions including:	
	(a) instructions to not remove or damage insulation when modifying landscaping.	
	(b) providing heat to the building as required by the ICC IRC or IBC.	
	(c) keeping base materials beneath and around the building free from moisture caused by broken water pipes or other water sources.	
(12)	A list of local service providers that offer regularly scheduled service and maintenance contracts to ensure proper performance of equipment and the structure (e.g., HVAC, water-heating equipment, sealants, caulks, gutter and downspout system, shower and/or tub surrounds, irrigation system).	
(13)	A photo record of framing with utilities installed. Photos are taken prior to installing insulation, clearly labeled, and included as part of the building owners' manual.	
(14)	List of common hazardous materials often used around the building and instructions for proper handling and disposal of these materials.	
(15)	Information on organic pest control, fertilizers, deicers, and cleaning products.	
(16)	Information on native landscape materials and/or those that have low-water requirements.	
(17)	Information on methods of maintaining the building's relative humidity in the range of 30 percent to 60 percent.	
(18)	Instructions for inspecting the building for termite infestation.	
(19)	Instructions for maintaining gutters and downspouts and importance of diverting water a minimum of 5 feet away from foundation.	
(20)	A narrative detailing the importance of maintenance and operation in retaining the attributes of a green-built building.	
(21)	Where stormwater management measures are installed on the lot, information on the location, purpose, and upkeep of these measures.	
(22)	For buildings originally built before 1978, the EPA publications "Reducing Lead Hazards When Remodeling Your Home" and "Abestos in Your Home: A Homeowner's Guide".	

11.1002
TRAINING OF BUILDING OWNERS ON OPERATION AND MAINTENANCE FOR ONE- AND TWO-FAMILY DWELLINGS AND MULTI-UNIT BUILDINGS

11.1002.1 Training of building owners. Building owners are familiarized with the role of occupants in achieving green goals. On-site training is provided to the responsible party(ies) regarding newly installed equipment operation and maintenance, control systems, and occupant actions that will improve the environmental performance of the building. These include:	Mandatory 8
(1) HVAC filters	
(2) thermostat operation and programming	
(3) lighting controls	

GREEN BUILDING PRACTICES	POINTS

(4)	appliances operation
(5)	water heater settings and hot water use
(6)	fan controls
(7)	recycling practices

11.1003
CONSTRUCTION, OPERATION, AND MAINTENANCE MANUALS AND TRAINING FOR MULTI-UNIT BUILDINGS

11.1003.0 Intent. Manuals are provided to the responsible parties (owner, management, tenant, and/or maintenance team) regarding the construction, operation, and maintenance of the building. Paper or digital format manuals are to include information regarding those aspects of the building's construction, maintenance, and operation that are within the area of responsibilities of the respective recipient. One or more responsible parties are to receive a copy of all documentation for archival purposes.

11.1003.1 Building construction manual. A building construction manual, including five or more of the following, is compiled and distributed in accordance with Section 11.1003.0.		1
(Points awarded per two items. Points awarded for both mandatory and non-mandatory items.)		
(1)	A narrative detailing the importance of constructing a green building, including a list of green building attributes included in the building. This narrative is included in all responsible parties' manuals.	**Mandatory**
(2)	A local green building program certificate as well as a copy of the *National Green Building Standard™*, as adopted by the Adopting Entity, and the individual measures achieved by the building.	**Mandatory**
(3)	Warranty, operation, and maintenance instructions for all equipment, fixtures, appliances, and finishes.	**Mandatory**
(4)	Record drawings of the building.	
(5)	A record drawing of the site including stormwater management plans, utility lines, landscaping with common name and genus/species of plantings.	
(6)	A diagram showing the location of safety valves and controls for major building systems.	
(7)	A list of the type and wattage of light bulbs installed in light fixtures.	
(8)	A photo record of framing with utilities installed. Photos are taken prior to installing insulation and clearly labeled.	

11.1003.2 Operations manual. Operations manuals are created and distributed to the responsible parties in accordance with Section 11.1003.0. Among all of the operation manuals, five or more of the following options are included.		1
(Points awarded per two items. Points awarded for both mandatory and non-mandatory items.)		
(1)	A narrative detailing the importance of operating and living in a green building. This narrative is included in all responsible parties' manuals.	**Mandatory**

GREEN BUILDING PRACTICES	POINTS
(2) A list of practices to conserve water and energy (e.g., turning off lights when not in use, switching the rotation of ceiling fans in changing seasons, purchasing ENERGY STAR appliances and electronics).	**Mandatory**
(3) Information on methods of maintaining the building's relative humidity in the range of 30 percent to 60 percent.	
(4) Information on opportunities to purchase renewable energy from local utilities or national green power providers and information on utility and tax incentives for the installation of on-site renewable energy systems.	
(5) Information on local and on-site recycling and hazardous waste disposal programs and, if applicable, building recycling and hazardous waste handling and disposal procedures.	
(6) Local public transportation options.	
(7) Explanation of the benefits of using compact fluorescent light bulbs, LEDs, or other high-efficiency lighting.	
(8) Information on native landscape materials and/or those that have low water requirements.	
(9) Information on the radon mitigation system, where applicable.	
(10) A procedure for educating tenants in rental properties on the proper use, benefits, and maintenance of green building systems including a maintenance staff notification process for improperly functioning equipment.	

GREEN BUILDING PRACTICES	POINTS
11.1003.3 Maintenance manual. Maintenance manuals are created and distributed to the responsible parties in accordance with Section 11.1003.0. Between all of the maintenance manuals, five or more of the following options are included. **(Points awarded per two items. Points awarded for both mandatory and non-mandatory items.)**	1
(1) A narrative detailing the importance of maintaining a green building. This narrative is included in all responsible parties' manuals.	**Mandatory**
(2) A list of local service providers that offer regularly scheduled service and maintenance contracts to ensure proper performance of equipment and the structure (e.g., HVAC, water-heating equipment, sealants, caulks, gutter and downspout system, shower and/or tub surrounds, irrigation system).	
(3) User-friendly maintenance checklist that includes:	
(a) HVAC filters	
(b) thermostat operation and programming	
(c) lighting controls	
(d) appliances and settings	
(e) water heater settings	
(f) fan controls	
(4) List of common hazardous materials often used around the building and instructions for proper handling and disposal of these materials.	
(5) Information on organic pest control, fertilizers, deicers, and cleaning products.	

GREEN BUILDING PRACTICES	POINTS

(6)	Instructions for maintaining gutters and downspouts and the importance of diverting water a minimum of 5 feet away from foundation.
(7)	Instructions for inspecting the building for termite infestation.
(8)	A procedure for rental tenant occupancy turnover that preserves the green features.
(9)	An outline of a formal green building training program for maintenance staff.

11.1004
INNOVATIVE PRACTICES

11.1004.1 (Reserved)

CHAPTER 12

REMODELING OF FUNCTIONAL AREAS

12.00
REMODELING OF FUNCTIONAL AREAS

12.0 Intent. This chapter sets forth the mandatory green building practices for remodeling functional areas of buildings. The intent of Chapter 12 is to address the most common remodeling projects: complete kitchen, full bathroom, complete basement, or an addition under 400 square feet. Chapter 12 is not intended to be used for rating minor alterations.

12.0.1 Applicability. Each applicable practice in Section 12.1 shall be met for any of the remodeled functional areas included in Chapter 12. Additionally, the requirements of Sections 12.2, 12.3, 12.4, or 12.5 that are specific to each of the functional areas shall be met. Unless otherwise required, the requirements of Chapter 12 only apply to the remodeled functional area.

12.1
GENERAL

12.1.601.2 Material usage. Structural systems required for the remodel are designed or construction techniques are implemented that reduce and optimize material usage using at least one of the following methods

(1)	Minimum structural member or element sizes necessary for strength and stiffness in accordance with advanced framing techniques or structural design standards are selected.
(2)	Higher-grade or higher-strength of the same materials than commonly specified for structural elements and components in the building are used and element or component sizes are reduced accordingly.
(3)	Performance-based structural design is used to optimize lateral force-resisting systems.

12.1.602.1.7.1 Moisture control measures. Moisture control measures for newly installed materials are in accordance with the following:

(1)	Building materials with visible mold are not installed or are cleaned or encapsulated prior to concealment and closing.
(2)	Insulation in cavities is dry in accordance with manufacturer's installation instructions when enclosed (e.g., with drywall).

12.1.602.1.7.2 Moisture content. Moisture content of subfloor, substrate, or concrete slabs is in accordance with the appropriate industry standard for the finish flooring to be applied during the remodel.

12.1.602.1.11 Tile backing materials. Newly installed tile backing materials under tiled surfaces in wet areas are in accordance with ASTM C1178, C1278, C1288, or C1325.

12.1(A) Product or material selection. At least two newly installed types of materials from Section 12.1(A) are used.

12.1(A).601.7 Site-applied finishing materials. One or more of the building materials or assemblies listed below that do not require additional site-applied material for finishing are incorporated in the remodel.

(a)	pigmented, stamped, decorative, or final finish concrete or masonry
(b)	interior trim not requiring paint or stain
(c)	exterior trim not requiring paint or stain
(d)	window, skylight, and door assemblies not requiring paint or stain on one of the following surfaces: 　i. exterior surfaces 　ii. interior surfaces
(e)	interior wall coverings or systems not requiring paint or stain or other type of finishing application
(f)	exterior wall coverings or systems not requiring paint or stain or other type of finishing application
(g)	pre-finished hardwood flooring

12.1(A).603.1 Reused and salvaged materials. Reclaimed and/or salvaged materials and components are used in the remodel.

12.1(A).604.1 Recycled content. Newly installed building materials with at least 25% recycled content are used for two components of the remodel.

12.1(A).606.1 Biobased products. Two or more of the following biobased products are used in the remodel.

(a)	certified solid wood in accordance with Section 12.1(A).606.2
(b)	engineered wood
(c)	bamboo
(d)	cotton
(e)	cork
(f)	straw
(g)	natural fiber products made from crops (soy-based, corn-based)
(h)	products with the minimum biobased contents of the USDA 7 CFR Part 2902
(i)	other biobased materials with a minimum of 50 percent biobased content (by weight or volume)

12.1(A).606.2 Wood-based products. Wood or wood-based products installed during the remodel are certified to the requirements of one of the following recognized product programs:

(a)	American Forest Foundation's American Tree Farm System® (ATFS)

GREEN BUILDING PRACTICES	

(b)	Canadian Standards Association's Sustainable Forest Management System Standards (CSA Z809)
(c)	Forest Stewardship Council (FSC)
(d)	Program for Endorsement of Forest Certification Systems (PEFC)
(e)	Sustainable Forestry Initiative® Program (SFI)
(f)	other product programs mutually recognized by PEFC

12.1(A).608.1 Resource-efficient materials. One or more products containing fewer materials are used in the remodel to achieve the same end-use requirements as conventional products, including but not limited to:

(a)	lighter, thinner brick with bed depth less than 3 inches and/or brick with coring of more that 25 percent
(b)	engineered wood or engineered steel products
(c)	roof or floor trusses

12.1(A).609.1 Regional materials. One or more regional materials are used in the remodel for major elements or components of the building.

12.1(A).610.1 Life cycle analysis. A life cycle analysis (LCA) tool is used to select environmentally preferable products or assemblies, or an LCA is conducted on the entire functional area in accordance with Section 12.1(A).610.1.1 or 12.1(A).610.1.2, respectively. Only one method of analysis or tool may be utilized. The reference service life is 60 years for any LCA tool. Results of the LCA are reported in terms of the environmental impacts listed in this practice and it is stated if operating energy was included in the LCA.

12.1(A).610.1.1 Functional area life cycle analysis. An LCA is performed for an entire functional area using a life cycle assessment and data compliant with ISO 14044 or other recognized standards.

12.1(A).610.1.2 Life cycle analysis for a product or assembly. An environmentally preferable product or assembly is selected for an application based upon the use of an LCA tool that incorporates data methods compliant with ISO 14044 or other recognized standards that compare the environmental impact of products or assemblies.

(1)	Two or more products with the same intended use are compared based on LCA and the product with at least a 15% average improvement is selected. A minimum of four environmental impact measures are included in the comparison. The environmental impact measures to be considered are chosen from the following:

	(a)	fossil fuel consumption
	(b)	global warming potential
	(c)	acidification potential
	(d)	eutrophication potential
	(e)	ozone depletion potential

(2) An assembly with improved environmental impact measures that are on average at least 15% better than a comparable functionally assembly is selected. A minimum of four environmental impact measures are included in the comparison. The full life cycle, from resource extraction to demolition and disposal (including but not limited to on-site construction, maintenance and replacement, material and product embodied acquisition, and process and transportation energy), is assessed. The assessment includes all structural elements, insulation, and wall coverings of the assembly. The assessment does not include electrical and mechanical equipment and controls, plumbing products, fire detection and alarm systems, elevators, and conveying systems. The following functional building elements are eligible for points under this practice:

(a) exterior walls

(b) roof/ceiling

(c) interior walls or ceilings

(d) intermediate floors

The environmental impact measures to be considered are chosen from the following:

(a) fossil fuel consumption

(b) global warming potential

(c) acidification potential

(d) eutrophication potential

(e) ozone depletion potential

12.1(A).611.1 Manufacturer's environmental management system concepts. For one or more products used in the remodel, the product's manufacturer's operations and business practices include environmental management system concepts, and the production facility is registered to ISO 14001 or equivalent.

12.1(A).611.2 Sustainable products. One or more of the following products are used. Certification third-party agency is ISO Guide 65 accredited.

(1) 50% or more of carpet installed (by square feet) is third-party certified to NSF/ANSI 140.

(2) 50% or more of resilient flooring installed (by square feet) is third-party certified to NSF/ANSI 332.

(3) 50% or more of the insulation installed (by square feet) is third-party certified to EcoLogo CCD-016.

(4) 50% or more of interior wall coverings installed (by square feet) is third-party certified to NSF/ANSI 342.

(5) 50% or more of the gypsum board installed (by square feet) is third-party certified to ULE ISR 100.

(6) 50% or more of the door leafs installed (by number of door leafs) is third-party certified to ULE ISR 102.

(7) 50% or more of the tile installed (by square feet) is third-party certified to ANSI A138.1 Specifications for sustainable ceramic tiles, glass tiles and tile installation materials.

GREEN BUILDING PRACTICES

12.1.605.0 Hazardous materials and waste. All hazardous materials exposed during the remodel are removed or comply with federal and local regulations. All waste classified as hazardous shall be properly handled and disposed of.

12.1.701.4.1.1 HVAC system sizing. Newly installed or modified space heating and cooling system is sized according to heating and cooling loads calculated using ACCA Manual J, or equivalent. New equipment is selected using ACCA Manual S or equivalent. Where existing equipment is used, Manual J is used to verify the capacity is appropriate for the remodel.

12.1.701.4.2.1 Duct air sealing. Newly installed or modified ducts or ducts that are exposed during the remodel are air sealed. All duct sealing materials are rated to UL 181A or UL 181B specifications and are used in accordance with manufacturer's instructions.

12.1.701.4.2.2 Supply ducts. Building cavities are not used as new supply ducts. Existing building cavities currently used as supply ducts exposed during the remodel are lined.

12.1.701.4.2.3 Duct system sizing. New duct system is sized and designed in accordance with ACCA Manual D or equivalent.

12.1.701.4.3.1 Building thermal envelope. The portions of the building thermal envelope that are exposed or created during the remodel are durably sealed to limit infiltration. The sealing methods between dissimilar materials allow for differential expansion and contraction. The following are caulked, gasketed, weather-stripped, or otherwise sealed with an air barrier material, suitable film, or solid material:

(a)	All joints, seams, and penetrations.
(b)	Site-built windows, doors, and skylights.
(c)	Openings between window and door assemblies and their respective jambs and framing.
(d)	Utility penetrations.
(e)	Dropped ceilings or chases adjacent to the thermal envelope.
(f)	Knee walls.
(g)	Walls and ceilings separating a garage from conditioned spaces.
(h)	Behind tubs and showers on exterior walls.
(i)	Common walls between dwelling units.
(j)	Attic access openings.
(k)	Rim joist junction.
(l)	Other sources of infiltration.

12.1.701.4.3.2 Air sealing and insulation. Grade 3 installation is not permitted for newly installed insulation. The compliance of the portions of the building envelope that are exposed or created during the remodel for air tightness and insulation installation is demonstrated via visual inspection. Building envelope tightness and insulation installation are considered acceptable when the items listed in Table 12.1.701.4.3.2(2) applicable to the method of construction are field verified.

Table 12.1.701.4.3.2(2)
Air Barrier and Insulation Inspection Component Criteria

COMPONENT	CRITERIA
Air barrier and thermal barrier	• Exterior thermal envelope insulation for framed walls is installed in substantial contact and continuous alignment with building envelope air barrier. • Breaks or joints in the air barrier are filled or repaired. • Air-permeable insulation is not used as a sealing material. • Air-permeable insulation is installed with an air barrier.
Ceiling/attic	• Air barrier in any dropped ceiling/soffit is substantially aligned with insulation and any gaps are sealed. • Attic access (except unvented attic), knee-wall door, or drop-down stair is sealed.
Walls	• Corners and headers are insulated. • Junction of foundation and sill plate is sealed.
Windows and doors	• Space between window/door jambs and framing is sealed.
Rim joists	• Rim joists are insulated and include an air barrier.
Floors (including above-garage and cantilevered floors)	• Insulation is installed to maintain permanent contact with underside of subfloor decking. • Air barrier is installed at any exposed edge of insulation.
Crawlspace walls	• Insulation is permanently attached to walls. • Exposed earth in unvented crawlspaces is covered with Class I vapor retarder with overlapping joints taped.
Shafts, penetrations	• Duct shafts, utility penetrations, knee walls and flue shafts opening to exterior or unconditioned space are sealed.
Narrow cavities	• Batts in narrow cavities are cut to fit, or narrow cavities are filled by sprayed/blown insulation.
Garage separation	• Air sealing is provided between the garage and conditioned spaces.
Recessed lighting	• Recessed light fixtures are air tight, IC rated, and sealed to drywall. • Exception—fixtures in conditioned space.
Plumbing and wiring	• Insulation is placed between outside and pipes. Batt insulation is cut to fit around wiring and plumbing, or sprayed/blown insulation extends behind piping and wiring.
Shower/tub on exterior wall	• Showers and tubs on exterior walls have insulation and an air barrier separating them from the exterior wall.
Electrical/phone box on exterior walls	• Air barrier extends behind boxes or air sealed-type boxes are installed.
Common wall	• Air barrier is installed in common wall between dwelling units.
HVAC register boots	• HVAC register boots that penetrate building envelope are sealed to subfloor or drywall.
Fireplace	• Fireplace walls include an air barrier.

GREEN BUILDING PRACTICES

12.1.701.4.3.3 Fenestration air leakage. Newly installed windows, skylights, and sliding glass doors (except site-built windows, skylights, and doors) have an air infiltration rate of no more than 0.3 cfm per square foot (1.5 L/s/m^2), and swinging doors no more than 0.5 cfm per square foot (2.6 L/s/m^2), when tested in accordance with NFRC 400 or AAMA/WDMA/CSA 101/I.S.2/A440 by an accredited, independent laboratory, and are listed and labeled.

12.1.701.4.3.4 Recessed lighting. Newly installed recessed luminaires installed in the building thermal envelope are sealed to limit air leakage between conditioned and unconditioned spaces. All recessed luminaires are IC-rated and labeled as meeting ASTM E 283 when tested at 1.57 psf (75 Pa) pressure differential with no more than 2.0 cfm (0.944 L/s) of air movement from the conditioned space to the ceiling cavity. All recessed luminaires are sealed with a gasket or caulk between the housing and the interior of the wall or ceiling covering.

12.1.701.4.4 High-efficacy lighting. A minimum of 50 percent of the hard-wired lighting fixtures and bulbs in the remodeled portion of the building, or the bulbs in those fixtures, qualify as high efficacy or equivalent.

12.1.701.4.5 Boiler supply piping. Newly installed boiler supply piping in unconditioned space that is accessible during the remodel is insulated.

12.1.703.5.3 Appliances. All newly installed major appliances in the remodeled portion of the building are ENERGY STAR or equivalent:

12.1.901.1.4 Gas-fired equipment. Newly installed gas-fired fireplaces and direct heating equipment is listed and is installed in accordance with the National Fuel Gas Code, International Fuel Gas Code, or the applicable local gas appliance installation code. Gas-fired fireplaces and direct heating equipment are vented to the outdoors.

12.1.901.2.1 Solid fuel-burning appliances. Newly installed solid fuel-burning fireplaces, inserts, stoves, and heaters are code compliant and are in accordance with the following requirements:

(1)   Site-built masonry wood-burning fireplaces are equipped with outside combustion air and a means of sealing the flue and the combustion air outlets to minimize interior air (heat) loss when not in operation.

(2)   Factory-built, wood-burning fireplaces are in accordance with the certification requirements of UL 127 and are EPA certified.

(3)   Wood stove and fireplace inserts, as defined in UL 1482 Section 3.8, are in accordance with the certification requirements of UL 1482 and are in accordance with the emission requirements of the EPA Certification and the State of Washington WAC 173-433-100(3).

   Pellet (biomass) stoves and furnaces are in accordance with the requirements of ASTM E1509 or are EPA certified

   Masonry heaters are in accordance with the definitions in ASTM E1602 and ICC IBC, Section 2112.1.

12.1.901.3 Attached garages. Newly installed door(s) in the common wall between the attached garage and conditioned space is tightly sealed and gasketed.

GREEN BUILDING PRACTICES

12.1.901.4 Wood materials. A minimum of 85 percent of newly installed wood structural panels is compliant with DOC PS 1 and/or DOC PS 2. OSB used for floor, wall, and/or roof sheathing is compliant with DOC PS 2. The panels are made with moisture-resistant adhesives. The trademark indicates these adhesives as follows: Exposure 1 or Exterior for plywood, and Exposure 1 for OSB.

12.1.901.5 Cabinets. A minimum of 85 percent of newly installed cabinets are in accordance with one or any combination of the following:

(1) All parts of the cabinet are made of solid wood or non-formaldehyde emitting materials such as metal or glass.

(2) The composite wood used in wood cabinets are in accordance with CARB Composite Wood Air Toxic Contaminant Measure Standard or equivalent as certified by a third-party program such as but not limited to, those in Appendix D.

12.1.901.6 Carpets. Carpets in the remodeled portion of the building are in accordance with the following:

(1) Wall-to-wall carpeting is not installed adjacent to water closets and bathing fixtures.

(2) A minimum of 85 percent of newly installed carpet area and carpet cushion (padding) are in accordance with the emission levels of CDPH/EHLB Standard Method v1.1, footnote b in Table 4.1 does not apply (i.e., maximum allowable formaldehyde concentration is 16.5 µg/m3 (13.5 ppb)). Emission levels are determined by a laboratory accredited to ISO/IEC 17025 and the CDPH/EHLB Standard Method v1.1 is in its scope of accreditation. The product is certified by a third-party program accredited to ISO Guide 65, such as, but not limited to, those in Appendix D.

12.1.901.7 Hard-surface flooring. At least 85 percent of all newly installed prefinished hard-surface flooring is in accordance with the emission concentration limits of CDPH/EHLB Standard Method v1.1, footnote b in Table 4.1 does not apply (i.e., maximum allowable formaldehyde concentration is 16.5 µg/m^3 (13.5 ppb)). Emission levels are determined by a laboratory accredited to ISO/IEC 17025 and the CDPH/EHLB Standard Method v1.1 is in its scope of accreditation. The product is certified by a third-party program accredited to ISO Guide 65, such as, but not limited to, those in Appendix D. Where post-manufacture coatings or surface applications have not been applied, the following hard surface flooring types are deemed to comply with the emission requirements of this section:

(a)	Ceramic tile flooring
(b)	Organic-free, mineral-based flooring
(c)	Clay masonry flooring
(d)	Concrete masonry flooring
(e)	Concrete flooring
(f)	Metal flooring
(g)	Glass

GREEN BUILDING PRACTICES

12.1.901.8 Interior wall coverings. At least 85 percent of newly installed interior wall coverings are in accordance with the emission concentration limits of CDPH/EHLB Standard Method v1.1, footnote b in Table 4.1 does not apply (i.e., maximum allowable formaldehyde concentration is 16.5 µg/m^3 (13.5 ppb)). Emission levels are determined by a laboratory accredited to ISO/IEC 17025 and the CDPH/EHLB Standard Method v1.1 is in its scope of accreditation. The product is certified by a third-party program accredited to ISO Guide 65, such as, but not limited to, those in Appendix D.

12.1.901.9 Architectural coatings. A minimum of 85 percent of newly applied architectural coatings in the remodeled portion of the building are in accordance with either Section 12.1.901.9.1 or Section 12.1.901.9.2.

12.1.901.9.1 New site-applied interior architectural coatings, which are inside the water-proofing envelope, are in accordance with one or more of the following:

(1) Zero VOC as determined by EPA Method 24 (VOC content below the detection limit for the method)

(2) GreenSeal GS-11 Standard for Paints and Coatings

(3) CARB Suggested Control Measure for Architectural Coatings (see Table 12.1.901.9.1).

Table 12.1.901.9.1
VOC Content Limits For Architectural Coatings[a,b,c]

Coating Category	LIMIT[d] (g/l)
Flat Coatings	50
Non-flat Coatings	100
Non-flat High-Gloss Coatings	150
Specialty Coatings:	
Aluminum Roof Coatings	400
Basement Specialty Coatings	400
Bituminous Roof Coatings	50
Bituminous Roof Primers	350
Bond Breakers	350
Concrete Curing Compounds	350
Concrete/Masonry Sealers	100
Driveway Sealers	50
Dry Fog Coatings	150
Faux Finishing Coatings	350
Fire Resistive Coatings	350
Floor Coatings	100
Form-Release Compounds	250
Graphic Arts Coatings (Sign Paints)	500
High Temperature Coatings	420

GREEN BUILDING PRACTICES

Coating Category	LIMIT[d] (g/l)
Industrial Maintenance Coatings	250
Low Solids Coatings	120[e]
Magnesite Cement Coatings	450
Mastic Texture Coatings	100
Metallic Pigmented Coatings	500
Multi-color Coatings	250
Pre-treatment Wash Primers	420
Primers, Sealers, and Undercoaters	100
Reactive Penetrating Sealers	350
Recycled Coatings	250
Roof Coatings	50
Rust Preventative Coatings	250
Shellacs, Clear	730
Shellacs, Opaque	550
Specialty Primers, Sealers, and Undercoaters	100
Stains	250
Stone Consolidants	450
Swimming Pool Coatings	340
Traffic Marking Coatings	100
Tub and Tile Refinish Coatings	420
Waterproofing Membranes	250
Wood Coatings	275
Wood Preservatives	350
Zinc-rich Primers	340

a. The specified limits remain in effect unless revised limits are listed in subsequent columns in the table.
b. Values in this table are derived from those specified by the California Air Resources Board, Architectural Coatings Suggested Control Measure, February 1, 2008.
c. Table 12.1.901.9.1 architectural coating regulatory category and VOC content compliance determination shall conform to the California Air Resources Board Suggested Control Measure for Architectural Coatings dated February 1, 2008.
d. Limits are expressed as VOC Regulatory (except as noted), thinned to the manufacturer's maximum thinning recommendation, excluding any colorant added to tint bases.
e. Limit is expressed as VOC actual.

12.1.901.9.2 New site-applied interior architectural coatings are in accordance with the emission levels of CDPH/EHLB Standard Method v1.1, footnote b in Table 4.1 does not apply (i.e., maximum allowable formaldehyde concentration is 16.5 $\mu g/m^3$ (13.5 ppb)). Emission levels are determined by a laboratory accredited to ISO/IEC 17025 and the CDPH/EHLB Standard Method v1.1 is in its scope of accreditation. The product is certified by a third-party program accredited to ISO Guide 65, such as, but not limited to, those in Appendix D.

GREEN BUILDING PRACTICES

12.1.901.10 Adhesives and sealants. Interior low-VOC adhesives and sealants located inside the waterproofing envelope: A minimum of 85 percent of newly applied site-applied adhesive and sealant products used within the interior of the building are in accordance with one of the following, as applicable.

(1) The emission levels of CDPH/EHLB Standard Method v1.1 when tested by a laboratory with the CDPH/EHLB Standard Method v1.1 within the laboratory scope of accreditation to ISO/IEC 17025 and certified by a third-party program accredited to ISO Guide 65, such as, but not limited to, those found in Appendix D.
Exception: Footnote b in Table 4.1 of CDPH/EHLB Standard Method v1.1 does not apply. Formaldehyde maximum allowable concentration is 16.5 µg/m3 (13.5 ppb).

(2) GreenSeal GS-36 Adhesives for Commercial Use

OR

(3) SCAQMD Rule 1168 (see Table 12.1.901.10.(3)), excluding products that are sold in 16-ounce containers or less and are regulated by the California Air Resource Board (CARB) Consumer Products Regulation.

Table 12.1.901.10.(3)
Site Applied Adhesive And Sealants Voc Limits[a,b]

ADHESIVE	VOC LIMIT (g/l)
Indoor carpet adhesives	50
Carpet pad adhesives	50
Outdoor carpet adhesives	150
Wood flooring adhesive	100
Rubber floor adhesives	60
Subfloor adhesives	50
Ceramic tile adhesives	65
VCT and asphalt tile adhesives	50
Drywall and panel adhesives	50
Cove base adhesives	50
Multipurpose construction adhesives	70
Structural glazing adhesives	100
Single ply roof membrane adhesives	250
Architectural sealants	250
Architectural sealant primer Non-porous Porous	 250 775
Modified bituminous sealant primer	500
Other sealant primers	750
CPVC solvent cement	490
PVC solvent cement	510
ABS solvent cement	325

GREEN BUILDING PRACTICES

ADHESIVE	VOC LIMIT (g/l)
Plastic cement welding	250
Adhesive primer for plastic	550
Contact adhesive	80
Special purpose contact adhesive	250
Structural wood member adhesive	140

a. VOC limit less water and less exempt compounds in grams/liter
b. For low-solid adhesives and sealants, the VOC limit is expressed in grams/liter of material as specified in Rule 1168. For all other adhesives and sealants, the VOC limits are expressed as grams of VOC per liter of adhesive or sealant less water and less exempt compounds as specified in Rule 1168.

12.1.901.11 Insulation. Emissions of newly installed wall, ceiling, and floor insulation materials are in accordance with the emission levels of CDPH/EHLB Standard Method v1.1, footnote b in Table 4.1 does not apply (i.e., maximum allowable formaldehyde concentration is 16.5 µg/m^3 (13.5 ppb)). Emission levels are determined by a laboratory accredited to ISO/IEC 17025 and the CDPH/EHLB Standard Method v1.1 is in its scope of accreditation. The product is certified by a third-party program accredited to ISO Guide 65, such as, but not limited to, those in Appendix D.

12.1.901.15 Lead-safe work practices. For buildings constructed before 1978, lead-safe work practices are used during the remodeling.

12.1.902.1.1 Spot ventilation. Spot ventilation is in accordance with the following:

(1) Bathrooms are vented to the outdoors. The minimum ventilation rate is 50 cfm (23.6 L/s) for intermittent operation or 20 cfm (9.4 L/s) for continuous operation in bathrooms.

(2) Clothes dryers are vented to the outdoors.

12.1.902.4 HVAC system protection. One of the following HVAC system protection measures is performed.

(1) HVAC supply registers (boots), return grilles, and rough-ins are covered during construction activities to prevent dust and other pollutants from entering the system.

(2) Prior to owner occupancy, HVAC supply registers (boots), return grilles, and duct terminations are inspected and vacuumed, if necessary. In addition, the coils are inspected and cleaned and the filter is replaced if necessary.

12.1.903.2 Duct insulation. All newly installed, exposed, or modified HVAC ducts, plenums, and trunks in unconditioned attics, basements, and crawlspaces are insulated to a minimum of R-6. Outdoor air supplies to ventilation systems are insulated to a minimum of R-6.

GREEN BUILDING PRACTICES

12.2
KITCHEN REMODELS

12.2.0 Applicability. In addition to the practices listed in Section 12.1, the following practices are mandatory for all kitchen remodels.

12.2.607.1 Recycling. Recycling by the occupants is facilitated by means of a built-in collection space in the kitchen or an aggregation/collection space in a garage, covered outdoor space, or other area for recycling containers.

12.2.611.4 Food waste disposers. Where allowed by local code, a food waste disposer is installed at each newly installed primary kitchen sink.

12.3
BATHROOM REMODELS

12.3.0 Applicability. In addition to the practices listed in Section 12.1, the following practices are mandatory for all bathroom remodels.

12.3.611.3 Universal design elements. Where existing walls are exposed and where new walls are constructed, blocking or equivalent is installed to accommodate the future installation of grab bars at water closet(s) and bathing fixture(s).

12.3.801.4 Showerheads. The total maximum combined flow rate of all newly installed showerheads that are controlled by a single valve at any point in time in a shower compartment is 1.6 to less than 2.5 gpm. Maximum of two valves are installed per shower compartment. The flow rate is tested at 80 psi (552 kPa) in accordance with ASME A112.18.1. Showerheads are served by an automatic compensating valve that complies with ASSE 1016 or ASME A112.18.1 and specifically designed to provide thermal shock and scald protection at the flow rate of the showerhead.

12.3.801.5.1 Faucets. Newly installed lavatory faucets have a maximum flow rate of 1.5 gpm (5.68 L/m) or less when tested at 60 psi (414 kPa) in accordance with ASME A112.18.1.

12.3.801.6 Water closets. All newly installed water closets have an effective flush volume of 1.28 gallons (4.85 L) or less when tested in accordance with ASME A112.19.2 or ASME A112.19.14 as applicable, and is in accordance with EPA WaterSense Tank-Type Toilets.

GREEN BUILDING PRACTICES

12.4 BASEMENT REMODELS

12.4.0 Applicability. In addition to the practices listed in Section 12.1, the following practices are mandatory for all basement remodels.

12.4.1 Moisture inspection. Prior to any construction activity, the basement is inspected for evidence of moisture problems. Any identified moisture problems are corrected prior to covering any walls or floors.

12.4.2 Kitchen. When the basement remodel includes a kitchen, the remodel shall also comply with the practices in Section 12.2.

12.4.3 Bathroom. When the basement remodel includes a bathroom, the remodel shall also comply with the practices in Section 12.3.

12.4.902.3 Radon control. In Radon Zone 1, passive or active radon control system is installed in accordance with ICC IRC Appendix F.

12.5 ADDITIONS

12.5.0 Applicability. In addition to the practices listed in Section 12.1, the following practices are mandatory for all addition remodels.

12.5.1 Kitchen. When the addition includes a kitchen, the remodel shall also comply with the practices in Section 12.2.

12.5.2 Bathroom. When the addition includes a bathroom, the remodel shall also comply with the practices in Section 12.3.

12.5.503.5 Landscape plan. Where the addition disturbs more than 1,000 square feet of the lot, a landscape plan for the lot is developed to limit water and energy use while preserving or enhancing the natural environment. Landscaping is phased to coincide with achievement of final grades to ensure denuded areas are quickly vegetated.

12.5.602.1.1.1 Capillary break. A capillary break and vapor retarder are installed at concrete slabs in the addition in accordance with IRC Sections R506.2.2 and R506.2.3 or IBC Sections 1910 and 1805.4.1.

12.5.602.1.3.1 Exterior drain tile. Where required by the ICC IRC or IBC for habitable and usable spaces of the addition below grade, exterior drain tile is installed.

GREEN BUILDING PRACTICES

12.5.602.1.4.1 Crawlspace. Vapor retarder in unconditioned vented crawlspace for the addition is in accordance with the following, as applicable. Joints of vapor retarder overlap a minimum of 6 inches (152 mm) and are taped.

(1) Floors. Minimum 6 mil vapor retarder installed on the crawlspace floor, extended at least 6 inches up the wall, and attached and sealed to the wall.

(2) Walls. Dampproof walls are provided below finished grade.

12.5.602.1.8 Water-resistive barrier. Where required by the ICC IRC or IBC, a water-resistive barrier and/or drainage-plane system is installed behind exterior veneer and/or siding of the addition.

12.5.602.1.9 Flashing. Flashing is provided for the addition and for the intersection where the addition joins the existing building, to minimize water entry into wall and roof assemblies and to direct water to exterior surfaces or exterior water-resistive barriers for drainage. Flashing details are provided in the construction documents and are in accordance with the fenestration manufacturer's instructions, the flashing manufacturer's instructions, or as detailed by a registered design professional. Flashing is installed at all of the following locations, as applicable:

(a) around exterior fenestrations, skylights, and doors

(b) at roof valleys

(c) at all building-to-deck, -balcony, -porch, and -stair intersections

(d) at roof-to-wall intersections, at roof-to-chimney intersections, at wall-to-chimney intersections, and at parapets.

(e) at ends of and under masonry, wood, or metal copings and sills

(f) above projecting wood trim

(g) at built-in roof gutters, and

(h) drip edge is installed at eaves and rake edges.

12.5.602.1.14 Ice barrier. In areas where there has been a history of ice forming along the eaves causing a backup of water, an ice barrier is installed on the addition in accordance with the ICC IRC or IBC at roof eaves of pitched roofs and extends a minimum of 24 inches (610 mm) inside the exterior wall line of the building.

12.5.602.1.15 Architectural features. New architectural features that increase the potential for water intrusion are avoided:

(1) No roof configurations that create horizontal valleys in roof design.

(2) No recessed windows and architectural features that trap water on horizontal surfaces.

(3) All horizontal ledgers are sloped away to provide gravity drainage as appropriate for the application.

GREEN BUILDING PRACTICES

12.5.602.4.1 Finished grade. Finished grade at all sides of the addition is sloped to provide a minimum of 6 inches (150 mm) of fall within 10 feet (3,048 mm) of the edge of the building. Where lot lines, walls, slopes, or other physical barriers prohibit 6 inches (152 mm) of fall within 10 feet (3,048 mm), the final grade is sloped away from the edge of the building at a minimum slope of 2 percent.

12.5.902.3 Radon control. In Radon Zone 1, a passive or active radon control system is installed in accordance with ICC IRC Appendix F.

CHAPTER 13

REFERENCED DOCUMENTS

1301 GENERAL

1301.1 This chapter lists the codes, standards, and other documents that are referenced in various sections of this Standard. The codes, standards, and other documents are listed herein indicating the promulgating agency of the document, the document identification, the effective date and title, and the section or sections of this Standard that reference the document. Unless indicated otherwise, the first printing of the document is referenced.

1301.2 The application of the referenced documents shall be as specified in Section 102.2.

1302 REFERENCED DOCUMENTS

ACCA		Air Conditioning Contractors of America 2800 Shirlington Road, Suite 300 Arlington, VA 22206 www.acca.org	(703) 575-4477
Manual D	2009	Residential Duct Systems	701.4.2.3, 11.701.4.2.3
Manual J	2006	Residential Load Calculation, Eighth Edition, Version 2	701.4.1.1, 701.4.1.2, 11.701.4.1.1, 11.701.4.1.2
Manual S	2004	Residential Equipment Selection	701.4.1.1, 11.701.4.1.1
5 QI	2010	HVAC Quality Installation Specification	701.4.1.2, 11.701.4.1.2

AFF		American Forest Foundation, Inc. 1111 Nineteenth Street, NW Suite 780 Washington, DC 20036 www.forestfoundation.org	(202) 463-2462
2010-2015 AFF Standards	2010	American Tree Farm System Standards for Sustainability for Forest Certification, including Performance Measures and Field Indicators	606.2(a), 11.606.2(a)

AAMA		American Architectural Manufacturers Association 1827 Walden Office Square, Suite 550 Schaumburg, Illinois 60173-4268 http://www.aamanet.org/	(847) 303-5664
711	2007	The Voluntary Specification for Self Adhering Flashing Used for Installation of Exterior Wall Fenestration Products	602.1.9(2), 11.602.1.9(2)
AAMA/WDMA/CSA 101/I.S.2/A440 UP3	2008		701.4.3.3, 11.701.4.3.3

AHRI		Air-Conditioning, Heating, and Refrigeration Institute (AHRI) 2111 Wilson Blvd, Suite 500 Arlington, VA 22201 www.ahrinet.org	(703) 524-8800
I=B=R	2009	Heat Loss Calculation Guide	701.4.1.2, 11.701.4.1.2

ASHRAE		American Society of Heating, Refrigerating, and Air-Conditioning Engineers, Inc. 1791 Tullie Circle, N.E. Atlanta, GA 30329 www.ashrae.org	(404) 636-8400
52.2	2007	Method of Testing General Ventilation Air Cleaning Devices for Removal Efficiency by Particle Size	202

ASCE		American Society of Civil Engineers 1801 Alexander Bell Drive Reston, VA 20191 www.asce.org	(800) 548-2723
32-01	2001	Design and Construction of Frost-Protected Shallow Foundations	202

ASME		American Society of Mechanical Engineers Three Park Avenue New York, NY 10016 www.asme.org	(800) 843-2763
A112.18.1	2005	Plumbing Supply Fittings	801.4(1), 801.5.1
A112.19.2/CSA B45.1	2008	Vitreous China Plumbing Fixtures and Hydraulic Requirements for Water Closets and Urinals	801.6(2), 801.6(3)
A112.19.14	2006	Six-Liter Water Closets Equipped with a Dual Flushing Device	801.6(2)

ASSE		American Society of Sanitary Engineering 901 Canterbury, Suite A Westlake, OH 44145 www.asse-plumbing.org	(440) 835-3040
1016	2011	Automatic Compensation Valves for Individual Showers and Tub/Shower Combinations	801.4(1)

ASTM		ASTM International, Inc. 100 Barr Harbor Drive, PO Box C700 West Conshohocken, PA 19428 www.astm.org	(610) 832-9500
C1178	2008	Standard Specification for Coated Glass Mat Water-Resistant Gypsum Backing Panel	602.1.11, 11.602.1.11
C1278 – 07a/1278M – 07a	2007	Standard Specification for Fiber-Reinforced Gypsum Panel	602.1.11, 11.602.1.11
C1288	2010	Standard Specification for Discrete Non-Asbestos Fiber-Cement Interior Substrate Sheets	602.1.11, 11.602.1.11
C1325	2008	Standard Specification for Non-Asbestos Fiber-Mat Reinforced Cement Substrate Sheets	602.1.11, 11.602.1.11
C1371	2010	Standard Test Method for Determination of Emittance of Materials Near Room Temperature Using Portable Emissometers	703.1.4
E283	2004	Standard Test Method for Determining Rate of Air Leakage Through Exterior Windows, Curtain Walls, and Doors Under Specified Pressure Differences Across the Specimen	701.4.3.4, 11.701.4.3.4
E1509	2005	Standard Specification for Room Heaters, Pellet Fuel-Burning Type	901.2.1(4), 11.901.2.1(4)
E1602	2010	Standard Guide for Construction of Solid Fuel Burning Masonry Heaters	901.2.1(5), 11.901.2.1(5)
E1980	2011	Standard Practice for Calculating Solar Reflectance Index of Horizontal and Low Sloped Opaque Surfaces	505.2(1),(2), 11.505.2(1),(2)

E2273	2011	Standard Test Method for Determining the Drainage Efficiency of Exterior Insulation and Finish Systems (EIFS) Clad Wall Assemblies	602.1.9(5)b, 11.602.1.9(5)b

CARB		*California Air Resources Board* *1001 "I" Street* *P.O. Box 2815* *Sacramento, CA 95812* *www.arb.ca.gov*	*(916) 322-2990*
	2007	Composite Wood Air Toxic Contaminant Measure Standard	901.4(5), 901.5(2), 11.901.4(5), 11.901.5(2)
	2008	Suggested Control Measure for Architectural Coatings	901.9.1(3), 11.901.9.1(3)
	2011	The California Consumer Products Regulations	901.10(3), 11.901.10(3)

CDPH		*California Department of Public Health* *850 Marina Bay Parkway* *Richmond, CA 94804* *www.cdph.ca.gov*	*(510) 620-2864*
	2010	Standard Method For The Testing And Evaluation Of Volatile Organic Chemical Emissions From Indoor Sources Using Environmental Chambers Version 1.1.	901.6(2), 901.7, 901.8, 901.9.3, 901.10(1), 901.11, 11.901.6(2), 11.901.7, 11.901.8, 11.901.9.3, 11.901.10(1), 11.901.11

CPA		*Composite Panel Association* *18922 Premiere Court* *Gaithersburg, MD 20879-1574* *www.pbmdf.com*	*(301) 670-0604*
A208.1	2009	Particleboard Standard	901.4(2), 11.901.4(2)
A208.2	2009	MDF Standard	901.4(2), 11.901.4(2)
CPA 4	20011	The Eco-Certified Composite™ (ECC) Standard	901.4(4), 11.901.4(4)

CSA		CSA International 8501 East Pleasant Valley Road Cleveland, OH 44131-5575 www.csa-international.org	(216) 524-4990
6.19	2006	Residential Carbon Monoxide Alarming Devices	901.12, 11.901.12
ANSI Z21.50/CSA 2.22	2007(2009 Addenda)	Vented Gas Fireplaces w/ Addenda b	901.1.5, 11.901.1.5
ANSI Z21.88/CSA 2.33	2009	Vented Gas Fireplace Heaters	901.1.5, 11.901.1.5
Z809	2008	Sustainable Forest Management Requirements and Guidance (SFM)	606.2(b), 11.606.2(b)

DOC/NIST		United States Department of Commerce National Institute of Standards and Technology 100 Bureau Drive Stop 3460 Gaithersburg, MD 20899-3460 www.nist.gov	(301) 975-2000
PS 1-09	2010	Construction and Industrial Plywood	901.4(1), 11.901.4(1)
PS 2-10	2011	Performance Standard for Wood-based Structural-use Panels	901.4(1), 11.901.4(1)

DOE		U.S. Department of Energy 1000 Independence Ave., SW Washington, DC 20585 www.energy.gov	800-345-3363
v. 4.4.2	2011	RESCheck	703.1.1

EcoLogo		The EcoLogo Program 171 Nepean Street, Suite 400 Ottawa, ON, K2P 0B4, CANADA	(800) 478-0399
CCD-016	2005	Thermal Insulation Materials	611.2(3), 11.611.2(3)

EPA		Environmental Protection Agency 1200 Pennsylvania Avenue, NW Washington, DC 20460 www.epa.gov	(202) 564-4700
EPA 747-K-97-001	1997	Reducing Lead Hazards When Remodeling Your Home	11.1001.1
Method 24	2000	Determination of Volatile Matter Content, Water Content, Density, Volume Solids, and Weight Solids of Surface Coatings	901.9.1(1), 11.901.9.1(1)
	1990	Asbestos in the Home: A Homeowner's Guide	11.1001.1
ENERGY STAR® Documents			
	August 29, 2011	ENERGY STAR for Homes Version 3.0 Guidelines	701.1.3
	January 1, 2011	ENERGY STAR Program Requirements for Clothes Washers, Version 5.1	703.5.3(3), 801.2(2), 801.2(3)
	January 20, 2012	ENERGY STAR Program Requirements for Dishwashers, Version 5.0	703.5.3(2), 801.2(1)
	December 1, 2009	ENERGY STAR Program Requirements for Geothermal Heat Pumps – Eligibility Criteria Version 3.1	703.2.6
	April 1, 2012	ENERGY STAR Program Requirements for Luminaires, Version 1.1	703.5.1(1)
	April 28, 2008	ENERGY STAR Program Eligibility Criteria for Residential Refrigerators and/or Freezers, Version 4.1	703.5.3(1)
	April 1, 2012	ENERGY STAR Program Requirements for Residential Ceiling Fans – Eligibility Criteria Version 3.0	703.2.7
	April 1, 2012	ENERGY STAR Program Requirements for Residential Ventilating Fans – Eligibility Criteria Version 3.2	902.1.4, 11.902.1.4
	January 1, 2011	ENERGY STAR Program Requirements for Residential Windows, Doors, and Skylights – Eligibility Criteria Version 5.0	703.6.1(3),(4), (5),(6)
	2010	ENERGY STAR Program Requirements for Roof Products – Eligibility Criteria Version 2.2	602.2(1), 11.602.2(1)
WaterSense Documents			
	May 20, 2011	WaterSense Specification for Tank-Type Toilets, Version 1.1	801.6(2)
	October 27, 2006	WaterSense: Professionals in System Design, Installation & Maintenance, and System Auditing	801.7.3
	December 9, 2009	WaterSense Water Budget Approach Version 1.0	403.6(4), 503.5(3), 11.503.5(3)

FSC		Forest Stewardship Council FSC International Center Charles-de-Gaulle 5 53113 Bonn, Germany *www.fsc.org*	*49 228 367 66 0*
FSC-STD-01-001 (Version 4-0) EN	2002	FSC Principles and Criteria for Forest Stewardship	606.2(c)

GS		Green Seal 1001 Connecticut Avenue, NW Suite 827 Washington, DC 20036 *www.greenseal.org*	*(202) 872-6400*
GS-11	2011	Paints and Coatings	901.9.1(2), 11.901.9.1(2)
GS-36	2011	Adhesives for Commercial Use	901.10(2), 11.901.10(2)

HPVA		Hardwood Plywood Veneer Association 1825 Michael Faraday Drive Reston, VA 20190 *www.hpva.org*	*(703) 435-2900*
HP-1	2009	American National Standard for Hardwood and Decorative Plywood	901.4(3), 11.901.4(3)

HUD		U.S. Department of Housing and Urban Development 451 7th Street SW Washington, DC 20410 *www.hud.gov*	*(202) 708-1112*
24 CFR, Part 3280	2005	Manufactured Home Construction and Safety Standards	202

ICC		International Code Council 500 New Jersey Ave, NW, 6th Floor Washington, DC 20001 _www.iccsafe.org_	(888) 422-7233
IBC	2009	International Building Code	202, 602.1.1.1, 602.1.3.1, 602.1.8, 602.1.14, 901.2.1(5), 1001.1(11)b
IECC	2004	International Energy Conservation Code	B201.1
IECC	2009	International Energy Conservation Code	701.1.1, 702.2.2
IFGC	2009	International Fuel Gas Code	901.1.4
IMC	2009	International Mechanical Code	704.5.1(1)
IRC	2009	International Residential Code	202, 602.1.1.1, 602.1.3.1, 602.1.8, 602.1.14, 704.5.1(1), 902.3, 1001.1(11)b

ISO		International Organization for Standardization 1, ch. de la Voie-Creuse, Case postale 56 CH-1211 Geneva 20, Switzerland _www.iso.org_	41 22 749 01 11
14044	2006	Environmental management – Life cycle assessment – Requirements and guidelines	610.1.1, 610.1.2, 11.610.1.1, 11.610.1.2
14001	2004	Environmental management systems – Requirements with guidance for use	611.1, 11.611.1
17025	2005	General requirements for the competence of testing and calibration laboratories	901.6(2), 901.7, 901.8, 901.9.3, 901.10(1), 901.11, 11.901.6(2), 11.901.7, 11.901.8, 11.901.9.3, 11.901.10(1), 11.901.11

Guide 65	1996	General requirements for bodies operating product certification systems	611.2, 901.6(2), 901.7, 901.8, 901.9.3, 901.10(1), 901.11, 11.611.2, 11.901.6(2), 11.901.7, 11.901.8, 11.901.9.3, 11.901.10(1), 11.901.11

NAHBRC		*NAHB Research Center* *400 Prince George's Boulevard* *Upper Marlboro, MD 20774* *www.nahbrc.org*	*(800) 638-8556*
Z765	2003	Single-Family Residential Buildings - Square Footage - Method for Calculating	601.1, 11.601.1

NFPA		*National Fire Protection Association* *1 Batterymarch Park* *Quincy, MA 02169* *www.nfpa.org*	*(617) 770-3000*
720	2012	Standard for the Installation of Carbon Monoxide (CO) Detection and Warning Equipment	901.12, 11.901.12
54	2012	National Fuel Gas Code	901.1.4, 11.901.1.4

NFRC		*National Fenestration Rating Council* *6305 Ivy Lane, Suite 140* *Greenbelt, MD 20770* *http://www.nfrc.org*	*(301) 589-1776*
400	2010	Procedure for Determining Fenestration Product Air Leakage	701.4.3.3, 11.701.4.3.3

NSF			NSF International P.O. Box 130140 789 N. Dixboro Road Ann Arbor, MI 48113-0140, USA www.nsf.org	(800) 673-6275
NSF/ANSI 140	2007	Sustainable Carpet Assessment		611.2(1), 11.611.2(1)
NSF/ANSI 332	2010	Sustainability Assessment for Resilient Floor Coverings		611.2(2), 11.611.2(2)
NSF/ANSI 342	2010	Sustainability Assessment for Wallcovering Products		611.2(4), 11.611.2(4)

PEFC			Pan European Forest Council 2éme Etage 17 Rue des Girondins Merl-Hollerich L - 1626 Luxembourg www.pefc.org	352 26 25 90 59
GL 2	2011	PEFC Council Minimum Requirements Checklist		606.2(d) & (f), 11.606.2(d) & (f)

SCAQMD			South Coast AQMD 21865 Copley Dr Diamond Bar, CA 91765	(909) 396-2000
Rule 1168	2005	Adhesive and Sealant Applications		901.10(3), 11.901.10(3)

SRCC			Solar Rating and Certification Corporation c/o FSEC 1679 Clearlake Road Cocoa, FL 32922-5703 www.solar-rating.org	(321) 638-1537
OG 300	2011	Operating Guidelines and Minimum Standards for Certifying Solar Water Heating Systems		703.4.5

SFI		Sustainable Forestry Initiative, Inc. 1600 Wilson Boulevard Suite 810 Arlington, VA 22209 www.sfiprogram.org	(703) 875-9500
2010-2014 Standard	2010	Sustainable Forestry Initiative Standard (SFIS)	606.2(e), 11.606.2(e)

TCIA		Tree Care Industry Association 3 Perimeter Road, Unit 1 Manchester, NH 03103 www.tcia.org	(603) 314-5380
A300	2001	Standards for Tree Care Operations - Tree, Shrub and Other Woody Plant Maintenance - Standard Practices	503.1(6), 11.503.1(6)

TCNA		Tile Council of North America 100 Clemson Research Blvd. Anderson, SC 29625 http://www.tileusa.com	(864) 646-8453
A138.1	2011	Green Squared: American National Standard Specifications for Sustainable Ceramic Tiles, Glass Tiles, and Tile Installation Materials	611.2(7), 11.611.2(7)

UL		Underwriters Laboratories Inc. 333 Pfingsten Road Northbrook, IL 60062-2096 www.ul.com	(877) 854-3577
127	2011	Factory-Built Fireplaces	901.2.1(2), 11.901.2.1(2)
181	2005	The Standard for Safety for Factory-Made Air Ducts and Air Connectors	701.4.2.1, 11.701.4.2.1
1482	2011	Solid-Fuel Type Room Heaters	901.2.1(3), 11.901.2.1(3)
2034	2007	Single and Multiple Station Carbon Monoxide Alarms	901.12, 11.901.12
100	2010	Interim Sustainability Requirements for Gypsum Boards and Panels	611.2(5), 11.611.2(5)
102	2012	Standard for Sustainability for Door Leafs	611.2(6), 11.611.2(6)

USDA		U.S. Department of Agriculture 1400 Independence Ave., SW Washington, DC 20250 *www.usda.gov*	*(202) 720-2791*
7 CFR Part 2902	2011	Designation of Biobased Items for Federal Procurement; Final Rule	606.1(h)

WSL		Washington State Legislature 106 Legislative Building Olympia, WA 98504-0600 *www.leg.wa.gov*	*(360) 786-7573*
WAC 173-433-100(3)	2007	Solid Fuel Burning Devices - Emission Performance Standards	901.2.1(3), 11.901.2.1(3)

APPENDIX A

DUCTED GARAGE EXHAUST FAN SIZING CRITERIA

A100
SCOPE AND APPLICABILITY

A101.1 Applicability of Appendix A. Appendix A is part of this Standard.

A101.2 Scope. The provisions contained in Appendix A provide the criteria necessary for complying with Section 901.3(1)(c) for the installation of ducted exhaust fans in garages. To receive points for implementing Practice 901.3(1)(c), the fan airflow rating and duct sizing for ducted exhaust fans are to be in accordance with the applicable criteria of Appendix A.

A101.3 Acknowledgment. The text of Appendix A, Section A200 and related Table are extracted from ASHRAE (American Society of Heating, Refrigerating and Air-Conditioning Engineers, Inc.) Standard 62.2-2007 *Ventilation and Acceptable Indoor Air Quality in Low-Rise Residential Buildings*, Section 7.3 and Table 7.1, respectively, and is used with the permission of ASHRAE. The referenced Section and Table numbers within the extracted text are modified to be applicable to Appendix A of this Standard.

A200
AIR FLOW RATING

A201.1 Airflow rating. The airflows required by this standard refer to the delivered airflow of the system as installed and tested using a flow hood, flow grid, or other airflow measuring device. Alternatively, the airflow rating at a pressure of 0.25 in. w.c. (62.5 Pa) may be used, provided the duct sizing meets the prescriptive requirements of Table A201 or manufacturers' design criteria.

TABLE A201
Prescriptive Duct Sizing

Fan Rating	Duct Type							
	Flex Duct				Smooth Duct			
cfm @ 0.25 in. w.g. (L/s @ 62.5 Pa)	50 (25)	80 (40)	100 (50)	125 (65)	50 (25)	80 (40)	100 (50)	125 (65)
Diameter, in. (mm)	Maximum Length, ft (m)							
3 (75)	X	X	X	X	5 (2)	X	X	X
4 (100)	70 (27)	3 (1)	X	X	105 (35)	35 (12)	5 (2)	X
5 (125)	NL	70 (27)	35 (12)	20 (7)	NL	135 (45)	85 (28)	55 (18)
6 (150)	NL	NL	125 (42)	95 (32)	NL	NL	NL	145 (48)
7 (175) and above	NL	NL	NL	NL	NL	NL	NL	NL

This table assumes no elbows. Deduct 15 ft (5 m) of allowable duct length for each elbow.

NL = no limit on duct length of this size.

X = not allowed, any length of duct of this size with assumed turns and fitting will exceed the rated pressure drop.

THIS PAGE INTENTIONALLY LEFT BLANK

APPENDIX B

WHOLE BUILDING VENTILATION SYSTEM SPECIFICATIONS

B100
SCOPE AND APPLICABILITY

B101.1 Applicability of Appendix B. Appendix B is part of this Standard.

B101.2 Scope. The provisions contained in Appendix B provide the specifications necessary for complying with Section 902.2.1 for the installation of whole building ventilation systems. To receive points for implementing Practice 902.2.1, the chosen whole building ventilation system is to be in accordance with the applicable specifications of Appendix B.

B101.3 Acknowledgment. The text of Appendix B, Section B200 and related Tables are extracted from ASHRAE (American Society of Heating, Refrigerating and Air-Conditioning Engineers, Inc.) Standard 62.2-2007 *Ventilation and Acceptable Indoor Air Quality in Low-Rise Residential Buildings*, Section 4, and is used with the permission of ASHRAE. The referenced Section and Table numbers within the extracted text are modified to be applicable to Appendix B of this Standard. "*" indicates added reference to ICC or ASHRAE 62.2 to provide clarity.

B200
WHOLE-BUILDING VENTILATION

B201.1 Ventilation Rate. A mechanical exhaust system, supply system, or combination thereof shall be installed for each dwelling unit to provide whole-building ventilation with outdoor air each hour at no less than the rate specified in Tables B201.1a and B201.1b or, equivalently, Equations B201.1a and B201.1b, based on the floor area of the conditioned space and number of bedrooms.

> **Exceptions:** Whole-building mechanical systems are not required provided that at least one of the following conditions is met:
>
> (a) the building is in zone 3B or 3C of the ICC* IECC 2004 Climate Zone Map (see ASHRAE 62.2*, Figure 8.2),
>
> (b) the building has no mechanical cooling and is in zone 1 or 2 of the ICC* IECC Climate Zone Map (see ASHRAE 62.2*, Figure 8.2), or
>
> (c) the building is thermally conditioned for human occupancy for less than 876 hours per year,
>
> *and* if the authority having jurisdiction determines that window operation is a locally permissible method of providing ventilation.

B201.1.1 Different Occupant Density. Tables B201.1a and B201.1b and Equations B201.1a and B201.1b assume two persons in a studio or one-bedroom dwelling unit and an additional person for each additional bedroom. Where higher occupant densities are known, the rate shall be increased by 7.5 cfm (3.5 L/s) for each additional person. When approved by the authority having jurisdiction, lower occupant densities may be used.

B201.1.2 Alternative Ventilation. Other methods may be used to provide the required ventilation rates (of Tables B201.1a and B201.1b) when approved by a licensed design professional.

B201.1.3 Infiltration Credit. Section B201.1 includes a default credit for ventilation provided by infiltration of 2 cfm/100 ft^2 (10 L/s per 100 m^2) of occupiable floor space. For buildings built prior to the application of this standard, when excess infiltration has been measured using *ANSI/ASHRAE Standard 136, A Method of Determining Air Change Rates in Detached Dwellings,*[1] the rates in Section B201.1 may be decreased by half of the excess of the rate calculated from Standard 136 that is above the default rate.

Equation B201.1a

$$Q_{fan} = 0.01A_{floor} + 7.5(N_{br} + 1)$$

where

Q_{fan} = fan flow rate, cfm

A_{floor} = floor area, ft^2

N_{br} = number of bedrooms; not to be less than one

Equation B201.1b

$$Q_{fan} = 0.05A_{floor} + 3.5(N_{br} + 1)$$

where

Q_{fan} = fan flow rate, L/s

A_{floor} = floor area, m^2

N_{br} = number of bedrooms; not to be less than one

TABLE B201.1a (I-P)
Ventilation Air Requirements, cfm

Floor Area (ft^2)	Bedrooms				
	0–1	2–3	4–5	6–7	>7
<1500	30	45	60	75	90
1501–3000	45	60	75	90	105
3001–4500	60	75	90	105	120
4501–6000	75	90	105	120	135
6001–7500	90	105	120	135	150
>7500	105	120	135	150	165

TABLE B201.1b (SI)
Ventilation Air Requirements, L/s

Floor Area (m^2)	Bedrooms				
	0–1	2–3	4–5	6–7	>7
<139	14	21	28	35	42
139.1–279	21	28	35	42	50
279.1–418	28	35	42	50	57
418.1–557	35	42	50	57	64
557.1–697	42	50	57	64	71
>697	50	57	64	71	78

---

[1] *ANSI/ASHRAE Standard 136-1993 (RA 2006), A Method of Determining Air Change Rates in Detached Dwellings.* American Society of Heating, Refrigerating and Air-Conditioning Engineers, Inc., Atlanta, GA.

B201.2 System Type. The whole-house ventilation system shall consist of one or more supply or exhaust fans and associated ducts and controls. Local exhaust fans shall be permitted to be part of a mechanical exhaust system. Outdoor air ducts connected to the return side of an air handler shall be permitted as supply ventilation if manufacturers' requirements for return air temperature are met. See ASHRAE 62.2*, Appendix B for guidance on selection of methods.

B201.3 Control and Operation. The "fan on" switch on a heating or air-conditioning system shall be permitted as an operational control for systems introducing ventilation air through a duct to the return side of an HVAC system. Readily accessible override control must be provided to the occupant. Local exhaust fan switches and "fan on" switches shall be permitted as override controls. Controls, including the "fan-on" switch of a conditioning system, must be appropriately labeled.

> **Exception:** An intermittently operating, whole-house mechanical ventilation system may be used if the ventilation rate is adjusted according to the exception to Section B201.4. The system must be designed so that it can operate automatically based on a timer. The intermittent mechanical ventilation system must operate at least one hour out of every twelve.

B201.4 Delivered Ventilation. The delivered ventilation rate shall be calculated as the larger of the total supply or total exhaust and shall be no less than specified in Section B201.1 during each hour of operation.

> **Exception:** The effective ventilation rate of an intermittent system is the combination of its delivered capacity, its daily fractional on-time, and the ventilation effectiveness from Table B201.2.

Equation B201.4

$$Q_f = Q_r / (\varepsilon f)$$

where

Q_f = fan flow rate

Q_r = ventilation air requirement (from Table B201.1a or B201.1b)

ε = ventilation effectiveness (from Table B201.2)

f = fractional on time

If the system runs at least once every three hours, 1.0 can be used as the ventilation effectiveness. (See *ASHRAE 62.2*, Appendix B for an example of this calculation.)

TABLE B201.4
Ventilation Effectiveness for Intermittent Fans

Daily Fractional On-Time, f	Ventilation Effectiveness, ε
$f \leq 35\%$	0.33
$35\% \leq f < 60\%$	0.50
$60\% \leq f < 80\%$	0.75
$80\% \leq f$	1.0

B201.5 Restrictions on System Type. Use of certain ventilation strategies is restricted in specific climates as follows.

B201.5.1 Hot, Humid Climates. In hot, humid climates, whole-house mechanical net exhaust flow shall not exceed 7.5 cfm per 100 ft^2 (35 L/s per 100 m^2). (See ASHRAE 62.2*, Section 8 for a listing of hot, humid US climates.)

B201.5.2 Very Cold Climates. Mechanical supply systems exceeding 7.5 cfm per 100 ft^2 (35 L/s per 100 m^2) shall not be used in very cold climates. (See ASHRAE 62.2*, Section 8 for a listing of very cold US climates.)

> **Exception:** These ventilation strategies are not restricted if the authority having jurisdiction approves the envelope design as being moisture resistant.

CLIMATE ZONES

C100
SCOPE AND APPLICABILITY

C101.1 Applicability of Appendix C. Appendix C is part of this Standard. Text identified as "User Note" is not considered part of this Standard.

C101.2 Scope. The provisions contained in Appendix C provide the criteria necessary for complying with the climate-specific provisions of this Standard.

C200
CLIMATE ZONES

TABLE C200
CLIMATE ZONES, MOISTURE REGIMES, AND WARM-HUMID DESIGNATIONS
BY STATE, COUNTY AND TERRITORY

Key: A – Moist, B – Dry, C – Marine. Absence of moisture designation indicates moisture regime is irrelevant.
Asterisk (*) indicates a warm-humid location.

ALABAMA
3A Autauga*
2A Baldwin*
3A Barbour*
3A Bibb
3A Blount
3A Bullock*
3A Butler*
3A Calhoun
3A Chambers
3A Cherokee
3A Chilton
3A Choctaw*
3A Clarke*
3A Clay
3A Cleburne
3A Coffee*
3A Colbert
3A Conecuh*
3A Coosa
3A Covington*
3A Crenshaw*
3A Cullman
3A Dale*
3A Dallas*
3A DeKalb

3A Elmore*
3A Escambia*
3A Etowah
3A Fayette
3A Franklin
3A Geneva*
3A Greene
3A Hale
3A Henry*
3A Houston*
3A Jackson
3A Jefferson
3A Lamar
3A Lauderdale
3A Lawrence
3A Lee
3A Limestone
3A Lowndes*
3A Macon*
3A Madison
3A Marengo*
3A Marion
3A Marshall
2A Mobile*
3A Monroe*
3A Montgomery*

3A Morgan
3A Perry*
3A Pickens
3A Pike*
3A Randolph
3A Russell*
3A Shelby
3A St. Clair
3A Sumter
3A Talladega
3A Tallapoosa
3A Tuscaloosa
3A Walker
3A Washington*
3A Wilcox*
3A Winston

ALASKA
7 Aleutians East
7 Aleutians West
7 Anchorage
8 Bethel
7 Bristol Bay
7 Denali
8 Dillingham
8 Fairbanks North Star

7 Haines
7 Juneau
7 Kenai Peninsula
7 Ketchikan Gateway
7 Kodiak Island
7 Lake and Peninsula
7 Matanuska-Susitna
8 Nome
8 North Slope
8 Northwest Arctic
7 Prince of Wales-Outer Ketchikan
7 Sitka
7 Skagway-Hoonah Angoon
8 Southeast Fairbanks
7 Valdez-Cordova
8 Wade Hampton
7 Wrangell-Petersburg
7 Yakutat
8 Yukon-Koyukuk

ARIZONA
5B Apache
3B Cochise
5B Coconino
4B Gila
3B Graham
3B Greenlee
2B La Paz
2B Maricopa
3B Mohave
5B Navajo
2B Pima
2B Pinal
3B Santa Cruz
4B Yavapai
4B Yuma

ARKANSAS
3A Arkansas
3A Ashley
4A Baxter
4A Benton
4A Boone
3A Bradley
3A Calhoun

(continued)

TABLE C200 – Continued
CLIMATE ZONES, MOISTURE REGIMES, AND WARM-HUMID DESIGNATIONS
BY STATE, COUNTY AND TERRITORY

Key: A – Moist, B – Dry, C – Marine. Absence of moisture designation indicates moisture regime is irrelevant.
Asterisk (*) indicates a warm-humid location.

4A Carroll	3A Perry	3C Marin	5B Boulder	6B Rio Blanco
3A Chicot	3A Phillips	4B Mariposa	6B Chaffee	7 Rio Grande
3A Clark	3A Pike	3C Mendocino	5B Cheyenne	7 Routt
3A Clay	3A Poinsett	3B Merced	7 Clear Creek	6B Saguache
3A Cleburne	3A Polk	5B Modoc	6B Conejos	7 San Juan
3A Cleveland	3A Pope	6B Mono	6B Costilla	6B San Miguel
3A Columbia*	3A Prairie	3C Monterey	5B Crowley	5B Sedgwick
3A Conway	3A Pulaski	3C Napa	6B Custer	7 Summit
3A Craighead	3A Randolph	5B Nevada	5B Delta	5B Teller
3A Crawford	3A Saline	3B Orange	5B Denver	5B Washington
3A Crittenden	3A Scott	3B Placer	6B Dolores	5B Weld
3A Cross	4A Searcy	5B Plumas	5B Douglas	5B Yuma
3A Dallas	3A Sebastian	3B Riverside	6B Eagle	
3A Desha	3A Sevier*	3B Sacramento	5B Elbert	**CONNECTICUT**
3A Drew	3A Sharp	3C San Benito	5B El Paso	5A (all)
3A Faulkner	3A St. Francis	3B San Bernardino	5B Fremont	
3A Franklin	4A Stone	3B San Diego	5B Garfield	**DELAWARE**
4A Fulton	3A Union*	3C San Francisco	5B Gilpin	4A (all)
3A Garland	3A Van Buren	3B San Joaquin	7 Grand	
3A Grant	4A Washington	3C San Luis Obispo	7 Gunnison	**DISTRICT OF**
3A Greene	3A White	3C San Mateo	7 Hinsdale	**COLUMBIA**
3A Hempstead*	3A Woodruff	3C Santa Barbara	5B Huerfano	4A (all)
3A Hot Spring	3A Yell	3C Santa Clara	7 Jackson	
3A Howard		3C Santa Cruz	5B Jefferson	**FLORIDA**
3A Independence	**CALIFORNIA**	3B Shasta	5B Kiowa	2A Alachua*
4A Izard	3C Alameda	5B Sierra	5B Kit Carson	2A Baker*
3A Jackson	6B Alpine	3B Siskiyou	7 Lake	2A Bay*
3A Jefferson	4B Amador	3B Solano	5B La Plata	2A Bradford*
3A Johnson	3B Butte	3C Sonoma	5B Larimer	2A Brevard*
3A Lafayette*	4B Calaveras	3B Stanislaus	4B Las Animas	1A Broward*
3A Lawrence	3B Colusa	3B Sutter	5B Lincoln	2A Calhoun*
3A Lee	3B Contra Costa	3B Tehama	5B Logan	2A Charlotte*
3A Lincoln	4C Del Norte	4B Trinity	5B Mesa	2A Citrus*
3A Little River*	4B El Dorado	3B Tulare	7 Mineral	2A Clay*
3A Logan	3B Fresno	4B Tuolumne	6B Moffat	2A Collier*
3A Lonoke	3B Glenn	3C Ventura	5B Montezuma	2A Columbia*
4A Madison	4C Humboldt	3B Yolo	5B Montrose	2A DeSoto*
4A Marion	2B Imperial	3B Yuba	5B Morgan	2A Dixie*
3A Miller*	4B Inyo		4B Otero	2A Duval*
3A Mississippi	3B Kern	**COLORADO**	6B Ouray	2A Escambia*
3A Monroe	3B Kings	5B Adams	7 Park	2A Flagler*
3A Montgomery	4B Lake	6B Alamosa	5B Phillips	2A Franklin*
3A Nevada	5B Lassen	5B Arapahoe	7 Pitkin	2A Gadsden*
4A Newton	3B Los Angeles	4B Baca	5B Prowers	2A Gilchrist*
3A Ouachita	3B Madera	5B Bent	5B Pueblo	2A Glades*

(continued)

ICC 700-2012 NATIONAL GREEN BUILDING STANDARD™

TABLE C200 – Continued
CLIMATE ZONES, MOISTURE REGIMES, AND WARM-HUMID DESIGNATIONS
BY STATE, COUNTY AND TERRITORY

Key: A – Moist, B – Dry, C – Marine. Absence of moisture designation indicates moisture regime is irrelevant.
Asterisk (*) indicates a warm-humid location.

2A Gulf*	2A Washington*	2A Decatur*	3A Lee*	3A Taylor*
2A Hamilton*		3A DeKalb	2A Liberty*	3A Telfair*
2A Hardee*	**GEORGIA**	3A Dodge*	3A Lincoln	3A Terrell*
2A Hendry*	2A Appling*	3A Dooly*	2A Long*	2A Thomas*
2A Hernando*	2A Atkinson*	3A Dougherty*	2A Lowndes*	3A Tift*
2A Highlands*	2A Bacon*	3A Douglas	4A Lumpkin	2A Toombs*
2A Hillsborough*	2A Baker*	3A Early*	3A Macon*	4A Towns
2A Holmes*	3A Baldwin	2A Echols*	3A Madison	3A Treutlen*
2A Indian River*	4A Banks	2A Effingham*	3A Marion*	3A Troup
2A Jackson*	3A Barrow	3A Elbert	3A McDuffie	3A Turner*
2A Jefferson*	3A Bartow	3A Emanuel*	2A McIntosh*	3A Twiggs*
2A Lafayette*	3A Ben Hill*	2A Evans*	3A Meriwether	4A Union
2A Lake*	2A Berrien*	4A Fannin	2A Miller*	3A Upson
2A Lee*	3A Bibb	3A Fayette	2A Mitchell*	4A Walker
2A Leon*	3A Bleckley*	4A Floyd	3A Monroe	3A Walton
2A Levy*	2A Brantley*	3A Forsyth	3A Montgomery*	2A Ware*
2A Liberty*	2A Brooks*	4A Franklin	3A Morgan	3A Warren
2A Madison*	2A Bryan*	3A Fulton	4A Murray	3A Washington
2A Manatee*	3A Bulloch*	4A Gilmer	3A Muscogee	2A Wayne*
2A Marion*	3A Burke	3A Glascock	3A Newton	3A Webster*
2A Martin*	3A Butts	2A Glynn*	3A Oconee	3A Wheeler*
1A Miami-Dade*	3A Calhoun*	4A Gordon	3A Oglethorpe	4A White
1A Monroe*	2A Camden*	2A Grady*	3A Paulding	4A Whitfield
2A Nassau*	3A Candler*	3A Greene	3A Peach*	3A Wilcox*
2A Okaloosa*	3A Carroll	3A Gwinnett	4A Pickens	3A Wilkes
2A Okeechobee*	4A Catoosa	4A Habersham	2A Pierce*	3A Wilkinson
2A Orange*	2A Charlton*	4A Hall	3A Pike	3A Worth*
2A Osceola*	2A Chatham*	3A Hancock	3A Polk	
2A Palm Beach*	3A Chattahoochee*	3A Haralson	3A Pulaski*	**HAWAII**
2A Pasco*	4A Chattooga	3A Harris	3A Putnam	1A (all)
2A Pinellas*	3A Cherokee	3A Hart	3A Quitman*	
2A Polk*	3A Clarke	3A Heard	4A Rabun	**IDAHO**
2A Putnam*	3A Clay*	3A Henry	3A Randolph*	5B Ada
2A Santa Rosa*	3A Clayton	3A Houston*	3A Richmond	6B Adams
2A Sarasota*	2A Clinch*	3A Irwin*	3A Rockdale	6B Bannock
2A Seminole*	3A Cobb	3A Jackson	3A Schley*	6B Bear Lake
2A St. Johns*	3A Coffee*	3A Jasper	3A Screven*	5B Benewah
2A St. Lucie*	2A Colquitt*	2A Jeff Davis*	2A Seminole*	6B Bingham
2A Sumter*	3A Columbia	3A Jefferson	3A Spalding	6B Blaine
2A Suwannee*	2A Cook*	3A Jenkins*	4A Stephens	6B Boise
2A Taylor*	3A Coweta	3A Johnson*	3A Stewart*	6B Bonner
2A Union*	3A Crawford	3A Jones	3A Sumter*	6B Bonneville
2A Volusia*	3A Crisp*	3A Lamar	3A Talbot	6B Boundary
2A Wakulla*	4A Dade	2A Lanier*	3A Taliaferro	6B Butte
2A Walton*	4A Dawson	3A Laurens*	2A Tattnall*	6B Camas

(continued)

TABLE C200 – Continued
CLIMATE ZONES, MOISTURE REGIMES, AND WARM-HUMID DESIGNATIONS
BY STATE, COUNTY AND TERRITORY

Key: A – Moist, B – Dry, C – Marine. Absence of moisture designation indicates moisture regime is irrelevant.
Asterisk (*) indicates a warm-humid location.

5B Canyon	4A Clay	4A Marion	**INDIANA**	5A Lake
6B Caribou	4A Clinton	5A Marshall	5A Adams	5A La Porte
5B Cassia	5A Coles	5A Mason	5A Allen	4A Lawrence
6B Clark	5A Cook	4A Massac	5A Bartholomew	5A Madison
5B Clearwater	4A Crawford	5A McDonough	5A Benton	5A Marion
6B Custer	5A Cumberland	5A McHenry	5A Blackford	5A Marshall
5B Elmore	5A DeKalb	5A McLean	5A Boone	4A Martin
6B Franklin	5A De Witt	5A Menard	4A Brown	5A Miami
6B Fremont	5A Douglas	5A Mercer	5A Carroll	4A Monroe
5B Gem	5A DuPage	4A Monroe	5A Cass	5A Montgomery
5B Gooding	5A Edgar	4A Montgomery	4A Clark	5A Morgan
5B Idaho	4A Edwards	5A Morgan	5A Clay	5A Newton
6B Jefferson	4A Effingham	5A Moultrie	5A Clinton	5A Noble
5B Jerome	4A Fayette	5A Ogle	4A Crawford	4A Ohio
5B Kootenai	5A Ford	5A Peoria	4A Daviess	4A Orange
5B Latah	4A Franklin	4A Perry	4A Dearborn	5A Owen
6B Lemhi	5A Fulton	5A Piatt	5A Decatur	5A Parke
5B Lewis	4A Gallatin	5A Pike	5A De Kalb	4A Perry
5B Lincoln	5A Greene	4A Pope	5A Delaware	4A Pike
6B Madison	5A Grundy	4A Pulaski	4A Dubois	5A Porter
5B Minidoka	4A Hamilton	5A Putnam	5A Elkhart	4A Posey
5B Nez Perce	5A Hancock	4A Randolph	5A Fayette	5A Pulaski
6B Oneida	4A Hardin	4A Richland	4A Floyd	5A Putnam
5B Owyhee	5A Henderson	5A Rock Island	5A Fountain	5A Randolph
5B Payette	5A Henry	4A Saline	5A Franklin	4A Ripley
5B Power	5A Iroquois	5A Sangamon	5A Fulton	5A Rush
5B Shoshone	4A Jackson	5A Schuyler	4A Gibson	4A Scott
6B Teton	4A Jasper	5A Scott	5A Grant	5A Shelby
5B Twin Falls	4A Jefferson	4A Shelby	4A Greene	4A Spencer
6B Valley	5A Jersey	5A Stark	5A Hamilton	5A Starke
5B Washington	5A Jo Daviess	4A St. Clair	5A Hancock	5A Steuben
	4A Johnson	5A Stephenson	4A Harrison	5A St. Joseph
ILLINOIS	5A Kane	5A Tazewell	5A Hendricks	4A Sullivan
5A Adams	5A Kankakee	4A Union	5A Henry	4A Switzerland
4A Alexander	5A Kendall	5A Vermilion	5A Howard	5A Tippecanoe
4A Bond	5A Knox	4A Wabash	5A Huntington	5A Tipton
5A Boone	5A Lake	5A Warren	4A Jackson	5A Union
5A Brown	5A La Salle	4A Washington	5A Jasper	4A Vanderburgh
5A Bureau	4A Lawrence	4A Wayne	5A Jay	5A Vermillion
5A Calhoun	5A Lee	4A White	4A Jefferson	5A Vigo
5A Carroll	5A Livingston	5A Whiteside	4A Jennings	5A Wabash
5A Cass	5A Logan	5A Will	5A Johnson	5A Warren
5A Champaign	5A Macon	4A Williamson	4A Knox	4A Warrick
4A Christian	4A Macoupin	5A Winnebago	5A Kosciusko	4A Washington
5A Clark	4A Madison	5A Woodford	5A Lagrange	5A Wayne

(continued)

ICC 700-2012 NATIONAL GREEN BUILDING STANDARD™

TABLE C200 – Continued
CLIMATE ZONES, MOISTURE REGIMES, AND WARM-HUMID DESIGNATIONS BY STATE, COUNTY AND TERRITORY

Key: A – Moist, B – Dry, C – Marine. Absence of moisture designation indicates moisture regime is irrelevant.
Asterisk (*) indicates a warm-humid location.

5A Wells	6A Hancock	5A Tama	4A Franklin	4A Pottawatomie
5A White	6A Hardin	5A Taylor	4A Geary	4A Pratt
5 Whitley	5A Harrison	5A Union	5A Gove	5A Rawlins
	5A Henry	5A Van Buren	5A Graham	4A Reno
IOWA	6A Howard	5A Wapello	4A Grant	5A Republic
5A Adair	6A Humboldt	5A Warren	4A Gray	4A Rice
5A Adams	6A Ida	5A Washington	5A Greeley	4A Riley
6A Allamakee	5A Iowa	5A Wayne	4A Greenwood	5A Rooks
5A Appanoose	5A Jackson	6A Webster	5A Hamilton	4A Rush
5A Audubon	5A Jasper	6A Winnebago	4A Harper	4A Russell
5A Benton	5A Jefferson	6A Winneshiek	4A Harvey	4A Saline
6A Black Hawk	5A Johnson	5A Woodbury	4A Haskell	5A Scott
5A Boone	5A Jones	6A Worth	4A Hodgeman	4A Sedgwick
6A Bremer	5A Keokuk	6A Wright	4A Jackson	4A Seward
6A Buchanan	6A Kossuth		4A Jefferson	4A Shawnee
6A Buena Vista	5A Lee	**KANSAS**	5A Jewell	5A Sheridan
6A Butler	5A Linn	4A Allen	4A Johnson	5A Sherman
6A Calhoun	5A Louisa	4A Anderson	4A Kearny	5A Smith
5A Carroll	5A Lucas	4A Atchison	4A Kingman	4A Stafford
5A Cass	6A Lyon	4A Barber	4A Kiowa	4A Stanton
5A Cedar	5A Madison	4A Barton	4A Labette	4A Stevens
6A Cerro Gordo	5A Mahaska	4A Bourbon	5A Lane	4A Sumner
6A Cherokee	5A Marion	4A Brown	4A Leavenworth	5A Thomas
6A Chickasaw	5A Marshall	4A Butler	4A Lincoln	5A Trego
5A Clarke	5A Mills	4A Chase	4A Linn	4A Wabaunsee
6A Clay	6A Mitchell	4A Chautauqua	5A Logan	5A Wallace
6A Clayton	5A Monona	4A Cherokee	4A Lyon	4A Washington
5A Clinton	5A Monroe	5A Cheyenne	4A Marion	5A Wichita
5A Crawford	5A Montgomery	4A Clark	4A Marshall	4A Wilson
5A Dallas	5A Muscatine	4A Clay	4A McPherson	4A Woodson
5A Davis	6A O'Brien	5A Cloud	4A Meade	4A Wyandotte
5A Decatur	6A Osceola	4A Coffey	4A Miami	
6A Delaware	5A Page	4A Comanche	5A Mitchell	**KENTUCKY**
5A Des Moines	6A Palo Alto	4A Cowley	4A Montgomery	4A (all)
6A Dickinson	6A Plymouth	4A Crawford	4A Morris	
5A Dubuque	6A Pocahontas	5A Decatur	4A Morton	**LOUISIANA**
6A Emmet	5A Polk	4A Dickinson	4A Nemaha	2A Acadia*
6A Fayette	5A Pottawattamie	4A Doniphan	4A Neosho	2A Allen*
6A Floyd	5A Poweshiek	4A Douglas	5A Ness	2A Ascension*
6A Franklin	5A Ringgold	4A Edwards	5A Norton	2A Assumption*
5A Fremont	6A Sac	4A Elk	4A Osage	2A Avoyelles*
5A Greene	5A Scott	5A Ellis	5A Osborne	2A Beauregard*
6A Grundy	5A Shelby	4A Ellsworth	4A Ottawa	3A Bienville*
5A Guthrie	6A Sioux	4A Finney	4A Pawnee	3A Bossier*
6A Hamilton	5A Story	4A Ford	5A Phillips	3A Caddo*

(continued)

TABLE C200 – Continued
CLIMATE ZONES, MOISTURE REGIMES, AND WARM-HUMID DESIGNATIONS
BY STATE, COUNTY AND TERRITORY

Key: A – Moist, B – Dry, C – Marine. Absence of moisture designation indicates moisture regime is irrelevant.
Asterisk (*) indicates a warm-humid location.

2A Calcasieu*
3A Caldwell*
2A Cameron*
3A Catahoula*
3A Claiborne*
3A Concordia*
3A De Soto*
2A East Baton Rouge*
3A East Carroll
2A East Feliciana*
2A Evangeline*
3A Franklin*
3A Grant*
2A Iberia*
2A Iberville*
3A Jackson*
2A Jefferson*
2A Jefferson Davis*
2A Lafayette*
2A Lafourche*
3A La Salle*
3A Lincoln*
2A Livingston*
3A Madison*
3A Morehouse
3A Natchitoches*
2A Orleans*
3A Ouachita*
2A Plaquemines*
2A Pointe Coupee*
2A Rapides*
3A Red River*
3A Richland*
3A Sabine*
2A St. Bernard*
2A St. Charles *
2A St. Helena*
2A St. James*
2A St. John the
    Baptist*
2A St. Landry*
2A St. Martin*
2A St. Mary*
2A St. Tammany*
2A Tangipahoa*

3A Tensas*
2A Terrebonne*
3A Union*
2A Vermilion*
3A Vernon*
2A Washington*
3A Webster*
2A West Baton
    Rouge*
3A West Carroll
2A West Feliciana*
3A Winn*

MAINE
6A Androscoggin
7 Aroostook
6A Cumberland
6A Franklin
6A Hancock
6A Kennebec
6A Knox
6A Lincoln
6A Oxford
6A Penobscot
6A Piscataquis
6A Sagadahoc
6A Somerset
6A Waldo
6A Washington
6A York

MARYLAND
4A Allegany
4A Anne Arundel
4A Baltimore
4A Baltimore (city)
4A Calvert
4A Caroline
4A Carroll
4A Cecil
4A Charles
4A Dorchester
4A Frederick
5A Garrett
4A Harford

4A Howard
4A Kent
4A Montgomery
4A Prince George's
4A Queen Anne's
4A Somerset
4A St. Mary's
4A Talbot
4A Washington
4A Wicomico
4A Worcester

MASSACHUSETTS
5A (all)

MICHIGAN
6A Alcona
6A Alger
5A Allegan
6A Alpena
6A Antrim
6A Arenac
7 Baraga
5A Barry
5A Bay
6A Benzie
5A Berrien
5A Branch
5A Calhoun
5A Cass
6A Charlevoix
6A Cheboygan
7 Chippewa
6A Clare
5A Clinton
6A Crawford
6A Delta
6A Dickinson
5A Eaton
6A Emmet
5A Genesee
6A Gladwin
7 Gogebic
6A Grand Traverse
5A Gratiot

5A Hillsdale
7 Houghton
6A Huron
5A Ingham
6A Iosco
7 Iron
6A Isabella
5A Jackson
5A Kalamazoo
6A Kalkaska
5A Kent
7 Keweenaw
6A Lake
5A Lapeer
6A Leelanau
5A Lenawee
5A Livingston
7 Luce
7 Mackinac
5A Macomb
6A Manistee
6A Marquette
6A Mason
6A Mecosta
6A Menominee
5A Midland
6A Missaukee
5A Monroe
5A Montcalm
6A Montmorency
5A Muskegon
6A Newaygo
5A Oakland
6A Oceana
6A Ogemaw
7 Ontonagon
6A Osceola
6A Oscoda
6A Otsego
5A Ottawa
6A Presque Isle
6A Roscommon
5A Saginaw
6A Sanilac

7 Schoolcraft
5A Shiawassee
5A St. Clair
5A St. Joseph
5A Tuscola
5A Van Buren
5A Washtenaw
5A Wayne
6A Wexford

MINNESOTA
7 Aitkin
6A Anoka
7 Becker
7 Beltrami
6A Benton
6A Big Stone
6A Blue Earth
6A Brown
7 Carlton
6A Carver
7 Cass
6A Chippewa
6A Chisago
7 Clay
7 Clearwater
7 Cook
6A Cottonwood
7 Crow Wing
6A Dakota
6A Dodge
6A Douglas
6A Faribault
6A Fillmore
6A Freeborn
6A Goodhue
7 Grant
6A Hennepin
6A Houston
7 Hubbard
6A Isanti
7 Itasca
6A Jackson
7 Kanabec
6A Kandiyohi

(continued)

ICC 700-2012 NATIONAL GREEN BUILDING STANDARD™

TABLE C200 – Continued
CLIMATE ZONES, MOISTURE REGIMES, AND WARM-HUMID DESIGNATIONS
BY STATE, COUNTY AND TERRITORY

Key: A – Moist, B – Dry, C – Marine. Absence of moisture designation indicates moisture regime is irrelevant.
Asterisk (*) indicates a warm-humid location.

7 Kittson	7 Wadena	3A Lafayette	3A Yalobusha	4A Henry
7 Koochiching	6A Waseca	3A Lamar*	3A Yazoo	4A Hickory
6A Lac qui Parle	6A Washington	3A Lauderdale		5A Holt
7 Lake	6A Watonwan	3A Lawrence*	**MISSOURI**	4A Howard
7 Lake of the Woods	7 Wilkin	3A Leake	5A Adair	4A Howell
6A Le Sueur	6A Winona	3A Lee	5A Andrew	4A Iron
6A Lincoln	6A Wright	3A Leflore	5A Atchison	4A Jackson
6A Lyon	6A Yellow Medicine	3A Lincoln*	4A Audrain	4A Jasper
7 Mahnomen		3A Lowndes	4A Barry	4A Jefferson
7 Marshall	**MISSISSIPPI**	3A Madison	4A Barton	4A Johnson
6A Martin	3A Adams*	3A Marion*	4A Bates	5A Knox
6A McLeod	3A Alcorn	3A Marshall	4A Benton	4A Laclede
6A Meeker	3A Amite*	3A Monroe	4A Bollinger	4A Lafayette
7 Mille Lacs	3A Attala	3A Montgomery	4A Boone	4A Lawrence
6A Morrison	3A Benton	3A Neshoba	5A Buchanan	5A Lewis
6A Mower	3A Bolivar	3A Newton	4A Butler	4A Lincoln
6A Murray	3A Calhoun	3A Noxubee	5A Caldwell	5A Linn
6A Nicollet	3A Carroll	3A Oktibbeha	4A Callaway	5A Livingston
6A Nobles	3A Chickasaw	3A Panola	4A Camden	5A Macon
7 Norman	3A Choctaw	2A Pearl River*	4A Cape Girardeau	4A Madison
6A Olmsted	3A Claiborne*	3A Perry*	4A Carroll	4A Maries
7 Otter Tail	3A Clarke	3A Pike*	4A Carter	5A Marion
7 Pennington	3A Clay	3A Pontotoc	4A Cass	4A McDonald
7 Pine	3A Coahoma	3A Prentiss	4A Cedar	5A Mercer
6A Pipestone	3A Copiah*	3A Quitman	5A Chariton	4A Miller
7 Polk	3A Covington*	3A Rankin*	4A Christian	4A Mississippi
6A Pope	3A DeSoto	3A Scott	5A Clark	4A Moniteau
6A Ramsey	3A Forrest*	3A Sharkey	4A Clay	4A Monroe
7 Red Lake	3A Franklin*	3A Simpson*	5A Clinton	4A Montgomery
6A Redwood	3A George*	3A Smith*	4A Cole	4A Morgan
6A Renville	3A Greene*	2A Stone*	4A Cooper	4A New Madrid
6A Rice	3A Grenada	3A Sunflower	4A Crawford	4A Newton
6A Rock	2A Hancock*	3A Tallahatchie	4A Dade	5A Nodaway
7 Roseau	2A Harrison*	3A Tate	4A Dallas	4A Oregon
6A Scott	3A Hinds*	3A Tippah	5A Daviess	4A Osage
6A Sherburne	3A Holmes	3A Tishomingo	5A DeKalb	4A Ozark
6A Sibley	3A Humphreys	3A Tunica	4A Dent	4A Pemiscot
6A Stearns	3A Issaquena	3A Union	4A Douglas	4A Perry
6A Steele	3A Itawamba	3A Walthall*	4A Dunklin	4A Pettis
6A Stevens	2A Jackson*	3A Warren*	4A Franklin	4A Phelps
7 St. Louis	3A Jasper	3A Washington	4A Gasconade	5A Pike
6A Swift	3A Jefferson*	3A Wayne*	5A Gentry	4A Platte
6A Todd	3A Jefferson Davis*	3A Webster	4A Greene	4A Polk
6A Traverse	3A Jones*	3A Wilkinson*	5A Grundy	5A Pulaski
6A Wabasha	3A Kemper	3A Winston	5A Harrison	5A Putnam

(continued)

TABLE C200 – Continued
CLIMATE ZONES, MOISTURE REGIMES, AND WARM-HUMID DESIGNATIONS
BY STATE, COUNTY AND TERRITORY

Key: A – Moist, B – Dry, C – Marine. Absence of moisture designation indicates moisture regime is irrelevant.
Asterisk (*) indicates a warm-humid location.

5A Ralls	5B Lander	**NEW MEXICO**	6A Clinton	6A Tompkins
4A Randolph	5B Lincoln	4B Bernalillo	5A Columbia	6A Ulster
4A Ray	5B Lyon	5B Catron	5A Cortland	6A Warren
4A Reynolds	5B Mineral	3B Chaves	6A Delaware	5A Washington
4A Ripley	5B Nye	4B Cibola	5A Dutchess	5A Wayne
4A Saline	5B Pershing	5B Colfax	5A Erie	4A Westchester
5A Schuyler	5B Storey	4B Curry	6A Essex	6A Wyoming
5A Scotland	5B Washoe	4B DeBaca	6A Franklin	5A Yates
4A Scott	5B White Pine	3B Dona Ana	6A Fulton	
4A Shannon		3B Eddy	5A Genesee	**NORTH**
5A Shelby	**NEW HAMPSHIRE**	4B Grant	5A Greene	**CAROLINA**
4A St. Charles	6A Belknap	4B Guadalupe	6A Hamilton	4A Alamance
4A St. Clair	6A Carroll	5B Harding	6A Herkimer	4A Alexander
4A Ste. Genevieve	5A Cheshire	3B Hidalgo	6A Jefferson	5A Alleghany
4A St. Francois	6A Coos	3B Lea	4A Kings	3A Anson
4A St. Louis	6A Grafton	4B Lincoln	6A Lewis	5A Ashe
4A St. Louis (city)	5A Hillsborough	5B Los Alamos	5A Livingston	5A Avery
4A Stoddard	6A Merrimack	3B Luna	6A Madison	3A Beaufort
4A Stone	5A Rockingham	5B McKinley	5A Monroe	4A Bertie
5A Sullivan	5A Strafford	5B Mora	6A Montgomery	3A Bladen
4A Taney	6A Sullivan	3B Otero	4A Nassau	3A Brunswick*
4A Texas		4B Quay	4A New York	4A Buncombe
4A Vernon	**NEW JERSEY**	5B Rio Arriba	5A Niagara	4A Burke
4A Warren	4A Atlantic	4B Roosevelt	6A Oneida	3A Cabarrus
4A Washington	5A Bergen	5B Sandoval	5A Onondaga	4A Caldwell
4A Wayne	4A Burlington	5B San Juan	5A Ontario	3A Camden
4A Webster	4A Camden	5B San Miguel	5A Orange	3A Carteret*
5A Worth	4A Cape May	5B Santa Fe	5A Orleans	4A Caswell
4A Wright	4A Cumberland	4B Sierra	5A Oswego	4A Catawba
	4A Essex	4B Socorro	6A Otsego	4A Chatham
MONTANA	4A Gloucester	5B Taos	5A Putnam	4A Cherokee
6B (all)	4A Hudson	5B Torrance	4A Queens	3A Chowan
	5A Hunterdon	4B Union	5A Rensselaer	4A Clay
NEBRASKA	5A Mercer	4B Valencia	4A Richmond	4A Cleveland
5A (all)	4A Middlesex		5A Rockland	3A Columbus*
	4A Monmouth	**NEW YORK**	5A Saratoga	3A Craven
NEVADA	5A Morris	5A Albany	5A Schenectady	3A Cumberland
5B Carson City (city)	4A Ocean	6A Allegany	6A Schoharie	3A Currituck
5B Churchill	5A Passaic	4A Bronx	6A Schuyler	3A Dare
3B Clark	4A Salem	6A Broome	5A Seneca	3A Davidson
5B Douglas	5A Somerset	6A Cattaraugus	6A Steuben	4A Davie
5B Elko	5A Sussex	5A Cayuga	6A St. Lawrence	3A Duplin
5B Esmeralda	4A Union	5A Chautauqua	4A Suffolk	4A Durham
5B Eureka	5A Warren	5A Chemung	6A Sullivan	3A Edgecombe
5B Humboldt		6A Chenango	5A Tioga	4A Forsyth

(continued)

ICC 700-2012 NATIONAL GREEN BUILDING STANDARD™

TABLE C200 – Continued
CLIMATE ZONES, MOISTURE REGIMES, AND WARM-HUMID DESIGNATIONS BY STATE, COUNTY AND TERRITORY

Key: A – Moist, B – Dry, C – Marine. Absence of moisture designation indicates moisture regime is irrelevant.
Asterisk (*) indicates a warm-humid location.

4A Franklin	3A Rowan	6A LaMoure	4A Clermont	5A Morgan
3A Gaston	4A Rutherford	6A Logan	5A Clinton	5A Morrow
4A Gates	3A Sampson	7 McHenry	5A Columbiana	5A Muskingum
4A Graham	3A Scotland	6A McIntosh	5A Coshocton	5A Noble
4A Granville	3A Stanly	6A McKenzie	5A Crawford	5A Ottawa
3A Greene	4A Stokes	7 McLean	5A Cuyahoga	5A Paulding
4A Guilford	4A Surry	6A Mercer	5A Darke	5A Perry
4A Halifax	4A Swain	6A Morton	5A Defiance	5A Pickaway
4A Harnett	4A Transylvania	7 Mountrail	5A Delaware	4A Pike
4A Haywood	3A Tyrrell	7 Nelson	5A Erie	5A Portage
4A Henderson	3A Union	6A Oliver	5A Fairfield	5A Preble
4A Hertford	4A Vance	7 Pembina	5A Fayette	5A Putnam
3A Hoke	4A Wake	7 Pierce	5A Franklin	5A Richland
3A Hyde	4A Warren	7 Ramsey	5A Fulton	5A Ross
4A Iredell	3A Washington	6A Ransom	4A Gallia	5A Sandusky
4A Jackson	5A Watauga	7 Renville	5A Geauga	4A Scioto
3A Johnston	3A Wayne	6A Richland	5A Greene	5A Seneca
3A Jones	4A Wilkes	7 Rolette	5A Guernsey	5A Shelby
4A Lee	3A Wilson	6A Sargent	4A Hamilton	5A Stark
3A Lenoir	4A Yadkin	7 Sheridan	5A Hancock	5A Summit
4A Lincoln	5A Yancey	6A Sioux	5A Hardin	5A Trumbull
4A Macon		6A Slope	5A Harrison	5A Tuscarawas
4A Madison	**NORTH DAKOTA**	6A Stark	5A Henry	5A Union
3A Martin	6A Adams	7 Steele	5A Highland	5A Van Wert
4A McDowell	7 Barnes	7 Stutsman	5A Hocking	5A Vinton
3A Mecklenburg	7 Benson	7 Towner	5A Holmes	5A Warren
5A Mitchell	6A Billings	7 Traill	5A Huron	4A Washington
3A Montgomery	7 Bottineau	7 Walsh	5A Jackson	5A Wayne
3A Moore	6A Bowman	7 Ward	5A Jefferson	5A Williams
4A Nash	7 Burke	7 Wells	5A Knox	5A Wood
3A New Hanover*	6A Burleigh	7 Williams	5A Lake	5A Wyandot
4A Northampton	7 Cass		4A Lawrence	
3A Onslow*	7 Cavalier	**OHIO**	5A Licking	**OKLAHOMA**
4A Orange	6A Dickey	4A Adams	5A Logan	3A Adair
3A Pamlico	7 Divide	5A Allen	5A Lorain	3A Alfalfa
3A Pasquotank	6A Dunn	5A Ashland	5A Lucas	3A Atoka
3A Pender*	7 Eddy	5A Ashtabula	5A Madison	4B Beaver
3A Perquimans	6A Emmons	5A Athens	5A Mahoning	3A Beckham
4A Person	7 Foster	5A Auglaize	5A Marion	3A Blaine
3A Pitt	6A Golden Valley	5A Belmont	5A Medina	3A Bryan
4A Polk	7 Grand Forks	4A Brown	5A Meigs	3A Caddo
3A Randolph	6A Grant	5A Butler	5A Mercer	3A Canadian
3A Richmond	7 Griggs	5A Carroll	5A Miami	3A Carter
3A Robeson	6A Hettinger	5A Champaign	5A Monroe	3A Cherokee
4A Rockingham	7 Kidder	5A Clark	5A Montgomery	3A Choctaw

(continued)

TABLE C200 – Continued
CLIMATE ZONES, MOISTURE REGIMES, AND WARM-HUMID DESIGNATIONS
BY STATE, COUNTY AND TERRITORY

Key: A – Moist, B – Dry, C – Marine. Absence of moisture designation indicates moisture regime is irrelevant.
Asterisk (*) indicates a warm-humid location.

4B Cimarron
3A Cleveland
3A Coal
3A Comanche
3A Cotton
3A Craig
3A Creek
3A Custer
3A Delaware
3A Dewey
3A Ellis
3A Garfield
3A Garvin
3A Grady
3A Grant
3A Greer
3A Harmon
3A Harper
3A Haskell
3A Hughes
3A Jackson
3A Jefferson
3A Johnston
3A Kay
3A Kingfisher
3A Kiowa
3A Latimer
3A Le Flore
3A Lincoln
3A Logan
3A Love
3A Major
3A Marshall
3A Mayes
3A McClain
3A McCurtain
3A McIntosh
3A Murray
3A Muskogee
3A Noble
3A Nowata
3A Okfuskee
3A Oklahoma
3A Okmulgee
3A Osage

3A Ottawa
3A Pawnee
3A Payne
3A Pittsburg
3A Pontotoc
3A Pottawatomie
3A Pushmataha
3A Roger Mills
3A Rogers
3A Seminole
3A Sequoyah
3A Stephens
4B Texas
3A Tillman
3A Tulsa
3A Wagoner
3A Washington
3A Washita
3A Woods
3A Woodward

OREGON
5B Baker
4C Benton
4C Clackamas
4C Clatsop
4C Columbia
4C Coos
5B Crook
4C Curry
5B Deschutes
4C Douglas
5B Gilliam
5B Grant
5B Harney
5B Hood River
4C Jackson
5B Jefferson
4C Josephine
5B Klamath
5B Lake
4C Lane
4C Lincoln
4C Linn
5B Malheur

4C Marion
5B Morrow
4C Multnomah
4C Polk
5B Sherman
4C Tillamook
5B Umatilla
5B Union
5B Wallowa
5B Wasco
4C Washington
5B Wheeler
4C Yamhill

PENNSYLVANIA
5A Adams
5A Allegheny
5A Armstrong
5A Beaver
5A Bedford
5A Berks
5A Blair
5A Bradford
4A Bucks
5A Butler
5A Cambria
6A Cameron
5A Carbon
5A Centre
4A Chester
5A Clarion
6A Clearfield
5A Clinton
5A Columbia
5A Crawford
5A Cumberland
5A Dauphin
4A Delaware
6A Elk
5A Erie
5A Fayette
5A Forest
5A Franklin
5A Fulton
5A Greene

5A Huntingdon
5A Indiana
5A Jefferson
5A Juniata
5A Lackawanna
5A Lancaster
5A Lawrence
5A Lebanon
5A Lehigh
5A Luzerne
5A Lycoming
6A McKean
5A Mercer
5A Mifflin
5A Monroe
4A Montgomery
5A Montour
5A Northampton
5A Northumberland
5A Perry
4A Philadelphia
5A Pike
6A Potter
5A Schuylkill
5A Snyder
5A Somerset
5A Sullivan
6A Susquehanna
6A Tioga
5A Union
5A Venango
5A Warren
5A Washington
6A Wayne
5A Westmoreland
5A Wyoming
4A York

RHODE ISLAND
5A (all)

SOUTH CAROLINA
3A Abbeville
3A Aiken

3A Allendale*
3A Anderson
3A Bamberg*
3A Barnwell*
3A Beaufort*
3A Berkeley*
3A Calhoun
3A Charleston*
3A Cherokee
3A Chester
3A Chesterfield
3A Clarendon
3A Colleton*
3A Darlington
3A Dillon
3A Dorchester*
3A Edgefield
3A Fairfield
3A Florence
3A Georgetown*
3A Greenville
3A Greenwood
3A Hampton*
3A Horry*
3A Jasper*
3A Kershaw
3A Lancaster
3A Laurens
3A Lee
3A Lexington
3A Marion
3A Marlboro
3A McCormick
3A Newberry
3A Oconee
3A Orangeburg
3A Pickens
3A Richland
3A Saluda
3A Spartanburg
3A Sumter
3A Union
3A Williamsburg
3A York

(continued)

ICC 700-2012 NATIONAL GREEN BUILDING STANDARD™

TABLE C200 – Continued
CLIMATE ZONES, MOISTURE REGIMES, AND WARM-HUMID DESIGNATIONS
BY STATE, COUNTY AND TERRITORY

Key: A – Moist, B – Dry, C – Marine. Absence of moisture designation indicates moisture regime is irrelevant.
Asterisk (*) indicates a warm-humid location.

SOUTH DAKOTA	6A McPherson	4A Dickson	4A Overton	2A Bexar*
6A Aurora	6A Meade	3A Dyer	4A Perry	3A Blanco*
6A Beadle	5A Mellette	3A Fayette	4A Pickett	3B Borden
5A Bennett	6A Miner	4A Fentress	4A Polk	2A Bosque*
5A Bon Homme	6A Minnehaha	4A Franklin	4A Putnam	3A Bowie*
6A Brookings	6A Moody	4A Gibson	4A Rhea	2A Brazoria*
6A Brown	6A Pennington	4A Giles	4A Roane	2A Brazos*
6A Brule	6A Perkins	4A Grainger	4A Robertson	3B Brewster
6A Buffalo	6A Potter	4A Greene	4A Rutherford	4B Briscoe
6A Butte	6A Roberts	4A Grundy	4A Scott	2A Brooks*
6A Campbell	6A Sanborn	4A Hamblen	4A Sequatchie	3A Brown*
5A Charles Mix	6A Shannon	4A Hamilton	4A Sevier	2A Burleson*
6A Clark	6A Spink	4A Hancock	3A Shelby	3A Burnet*
5A Clay	6A Stanley	3A Hardeman	4A Smith	2A Caldwell*
6A Codington	6A Sully	3A Hardin	4A Stewart	2A Calhoun*
6A Corson	5A Todd	4A Hawkins	4A Sullivan	3B Callahan
6A Custer	5A Tripp	3A Haywood	4A Sumner	2A Cameron*
6A Davison	6A Turner	3A Henderson	3A Tipton	3A Camp*
6A Day	5A Union	4A Henry	4A Trousdale	4B Carson
6A Deuel	6A Walworth	4A Hickman	4A Unicoi	3A Cass*
6A Dewey	5A Yankton	4A Houston	4A Union	4B Castro
5A Douglas	6A Ziebach	4A Humphreys	4A Van Buren	2A Chambers*
6A Edmunds		4A Jackson	4A Warren	2A Cherokee*
6A Fall River	**TENNESSEE**	4A Jefferson	4A Washington	3B Childress
6A Faulk	4A Anderson	4A Johnson	4A Wayne	3A Clay
6A Grant	4A Bedford	4A Knox	4A Weakley	4B Cochran
5A Gregory	4A Benton	3A Lake	4A White	3B Coke
6A Haakon	4A Bledsoe	3A Lauderdale	4A Williamson	3B Coleman
6A Hamlin	4A Blount	4A Lawrence	4A Wilson	3A Collin*
6A Hand	4A Bradley	4A Lewis		3B Collingsworth
6A Hanson	4A Campbell	4A Lincoln	**TEXAS**	2A Colorado*
6A Harding	4A Cannon	4A Loudon	2A Anderson*	2A Comal*
6A Hughes	4A Carroll	4A Macon	3B Andrews	3A Comanche*
5A Hutchinson	4A Carter	4A Madison	2A Angelina*	3B Concho
6A Hyde	4A Cheatham	4A Marion	2A Aransas*	3A Cooke
5A Jackson	3A Chester	4A Marshall	3A Archer	2A Coryell*
6A Jerauld	4A Claiborne	4A Maury	4B Armstrong	3B Cottle
6A Jones	4A Clay	4A McMinn	2A Atascosa*	3B Crane
6A Kingsbury	4A Cocke	3A McNairy	2A Austin*	3B Crockett
6A Lake	4A Coffee	4A Meigs	4B Bailey	3B Crosby
6A Lawrence	3A Crockett	4A Monroe	2B Bandera*	3B Culberson
6A Lincoln	4A Cumberland	4A Montgomery	2A Bastrop*	4B Dallam
6A Lyman	4A Davidson	4A Moore	3B Baylor	3A Dallas*
6A Marshall	4A Decatur	4A Morgan	2A Bee*	3B Dawson
6A McCook	4A DeKalb	4A Obion	2A Bell*	4B Deaf Smith

(continued)

TABLE C200 – Continued
CLIMATE ZONES, MOISTURE REGIMES, AND WARM-HUMID DESIGNATIONS
BY STATE, COUNTY AND TERRITORY

Key: A – Moist, B – Dry, C – Marine. Absence of moisture designation indicates moisture regime is irrelevant.
Asterisk (*) indicates a warm-humid location.

3A Delta	2A Hays*	3A Llano*	3B Reeves	2B Webb*
3A Denton*	3B Hemphill	3B Loving	2A Refugio*	2A Wharton*
2A DeWitt*	3A Henderson*	3B Lubbock	4B Roberts	3B Wheeler
3B Dickens	2A Hidalgo*	3B Lynn	2A Robertson*	3A Wichita
2B Dimmit*	2A Hill*	2A Madison*	3A Rockwall*	3B Wilbarger
4B Donley	4B Hockley	3A Marion*	3B Runnels	2A Willacy*
2A Duval*	3A Hood*	3B Martin	3A Rusk*	2A Williamson*
3A Eastland	3A Hopkins*	3B Mason	3A Sabine*	2A Wilson*
3B Ector	2A Houston*	2A Matagorda*	3A San Augustine*	3B Winkler
2B Edwards*	3B Howard	2B Maverick*	2A San Jacinto*	3A Wise
3A Ellis*	3B Hudspeth	3B McCulloch	2A San Patricio*	3A Wood*
3B El Paso	3A Hunt*	2A McLennan*	3A San Saba*	4B Yoakum
3A Erath*	4B Hutchinson	2A McMullen*	3B Schleicher	3A Young
2A Falls*	3B Irion	2B Medina*	3B Scurry	2B Zapata*
3A Fannin	3A Jack	3B Menard	3B Shackelford	2B Zavala*
2A Fayette*	2A Jackson*	3B Midland	3A Shelby*	
3B Fisher	2A Jasper*	2A Milam*	4B Sherman	**UTAH**
4B Floyd	3B Jeff Davis	3A Mills*	3A Smith*	5B Beaver
3B Foard	2A Jefferson*	3B Mitchell	3A Somervell*	6B Box Elder
2A Fort Bend*	2A Jim Hogg*	3A Montague	2A Starr*	6B Cache
3A Franklin*	2A Jim Wells*	2A Montgomery*	3A Stephens	6B Carbon
2A Freestone*	3A Johnson*	4B Moore	3B Sterling	6B Daggett
2B Frio*	3B Jones	3A Morris*	3B Stonewall	5B Davis
3B Gaines	2A Karnes*	3B Motley	3B Sutton	6B Duchesne
2A Galveston*	3A Kaufman*	3A Nacogdoches*	4B Swisher	5B Emery
3B Garza	3A Kendall*	3A Navarro*	3A Tarrant*	5B Garfield
3A Gillespie*	2A Kenedy*	2A Newton*	3B Taylor	5B Grand
3B Glasscock	3B Kent	3B Nolan	3B Terrell	5B Iron
2A Goliad*	3B Kerr	2A Nueces*	3B Terry	5B Juab
2A Gonzales*	3B Kimble	4B Ochiltree	3B Throckmorton	5B Kane
4B Gray	3B King	4B Oldham	3A Titus*	5B Millard
3A Grayson	2B Kinney*	2A Orange*	3B Tom Green	6B Morgan
3A Gregg*	2A Kleberg*	3A Palo Pinto*	2A Travis*	5B Piute
2A Grimes*	3B Knox	3A Panola*	2A Trinity*	6B Rich
2A Guadalupe*	3A Lamar*	3A Parker*	2A Tyler*	5B Salt Lake
4B Hale	4B Lamb	4B Parmer	3A Upshur*	5B San Juan
3B Hall	3A Lampasas*	3B Pecos	3B Upton	5B Sanpete
3A Hamilton*	2B La Salle*	2A Polk*	2B Uvalde*	5B Sevier
4B Hansford	2A Lavaca*	4B Potter	2B Val Verde*	6B Summit
3B Hardeman	2A Lee*	3B Presidio	3A Van Zandt*	5B Tooele
2A Hardin*	2A Leon*	3A Rains*	2A Victoria*	6B Uintah
2A Harris*	2A Liberty*	4B Randall	2A Walker*	5B Utah
3A Harrison*	2A Limestone*	3B Reagan	2A Waller*	6B Wasatch
4B Hartley	4B Lipscomb	2B Real*	3B Ward	3B Washington
3B Haskell	2A Live Oak*	3A Red River*	2A Washington*	5B Wayne

(continued)

TABLE C200 – Continued
CLIMATE ZONES, MOISTURE REGIMES, AND WARM-HUMID DESIGNATIONS BY STATE, COUNTY AND TERRITORY

Key: A – Moist, B – Dry, C – Marine. Absence of moisture designation indicates moisture regime is irrelevant.
Asterisk (*) indicates a warm-humid location.

5B Weber

VERMONT

6A (all)

VIRGINIA

4A (all)

WASHINGTON

5B Adams
5B Asotin
5B Benton
5B Chelan
4C Clallam
4C Clark
5B Columbia
4C Cowlitz
5B Douglas
6B Ferry
5B Franklin
5B Garfield
5B Grant
4C Grays Harbor
4C Island
4C Jefferson
4C King
4C Kitsap
5B Kittitas
5B Klickitat
4C Lewis
5B Lincoln
4C Mason
6B Okanogan
4C Pacific
6B Pend Oreille
4C Pierce
4C San Juan
4C Skagit
5B Skamania
4C Snohomish
5B Spokane
6B Stevens
4C Thurston
4C Wahkiakum
5B Walla Walla

4C Whatcom
5B Whitman
5B Yakima

WEST VIRGINIA

5A Barbour
4A Berkeley
4A Boone
4A Braxton
5A Brooke
4A Cabell
4A Calhoun
4A Clay
5A Doddridge
5A Fayette
4A Gilmer
5A Grant
5A Greenbrier
5A Hampshire
5A Hancock
5A Hardy
5A Harrison
4A Jackson
4A Jefferson
4A Kanawha
5A Lewis
4A Lincoln
4A Logan
5A Marion
5A Marshall
4A Mason
4A McDowell
4A Mercer
5A Mineral
4A Mingo
5A Monongalia
4A Monroe
4A Morgan
5A Nicholas
5A Ohio
5A Pendleton
4A Pleasants
5A Pocahontas
5A Preston
4A Putna

5A Raleigh
5A Randolph
4A Ritchie
4A Roane
5A Summers
5A Taylor
5A Tucker
4A Tyler
5A Upshur
4A Wayne
5A Webster
5A Wetzel
4A Wirt
4A Wood
4A Wyoming

WISCONSIN

6A Adams
7 Ashland
6A Barron
7 Bayfield
6A Brown
6A Buffalo
7 Burnett
6A Calumet
6A Chippewa
6A Clark
6A Columbia
6A Crawford
6A Dane
6A Dodge
6A Door
7 Douglas
6A Dunn
6A Eau Claire
7 Florence
6A Fond du Lac
7 Forest
6A Grant
6A Green
6A Green Lake
6A Iowa
7 Iron
6A Jackson
6A Jefferson

6A Juneau
6A Kenosha
6A Kewaunee
6A La Crosse
6A Lafayette
7 Langlade
7 Lincoln
6A Manitowoc
6A Marathon
6A Marinette
6A Marquette
6A Menominee
6A Milwaukee
6A Monroe
6A Oconto
7 Oneida
6A Outagamie
6A Ozaukee
6A Pepin
6A Pierce
6A Polk
6A Portage
7 Price
6A Racine
6A Richland
6A Rock
6A Rusk
6A Sauk
7 Sawyer
6A Shawano
6A Sheboygan
6A St. Croix
7 Taylor
6A Trempealeau
6A Vernon
7 Vilas
6A Walworth
7 Washburn
6A Washington
6A Waukesha
6A Waupaca
6A Waushara
6A Winnebago
6A Wood

WYOMING

6B Albany
6B Big Horn
6B Campbell
6B Carbon
6B Converse
6B Crook
6B Fremont
5B Goshen
6B Hot Springs
6B Johnson
6B Laramie
7 Lincoln
6B Natrona
6B Niobrara
6B Park
5B Platte
6B Sheridan
7 Sublette
6B Sweetwater
7 Teton
6B Uinta
6B Washakie
6B Weston

U.S. TERRITORIES

AMERICAN SAMOA

1A (all)*

GUAM

1A (all)*

NORTHERN MARIANA ISLANDS

1A (all)*

PUERTO RICO

1A (all)*

VIRGIN ISLANDS

1A (all)*

C300
INTERNATIONAL CLIMATE ZONES

C301 International climate zones. The climate *zone* for any location outside the United States shall be determined by applying Table C301(1) and then Table C301(2).

TABLE C301(1)
INTERNATIONAL CLIMATE ZONE DEFINITIONS

MAJOR CLIMATE TYPE DEFINITIONS
Marine (C) Definition – Locations meeting all four criteria: 1. Mean temperature of coldest month between -3°C (27°F) and 18°C (65°F) 2. Warmest month mean <22°C (72°F) 3. At least four months with mean temperatures over 10°C (50°F) 4. Dry season in summer. The month with the heaviest precipitation in the cold season has at least three times as much precipitation as the month with the least precipitation in the rest of the year. The cold season is October through March in the Northern Hemisphere and April through September in the Southern Hemisphere.
Dry (B) Definition—Locations meeting the following criteria: Not marine and P_{in} <0.44 × (*TF* - 19.5)   [P_{cm}<2.0 × (*TC* + 7) in SI units] where: P_{in} = Annual precipitation in inches (cm) *T* = Annual mean temperature in °F (°C)
Moist (A) Definition – Locations that are not marine and not dry.
Warm-humid Definition – Moist (A) locations where either of the following wet-bulb temperature conditions shall occur during the warmest six consecutive months of the year: 1. 67°F (19.4°C) or higher for 3,000 or more hours; or 2. 73°F (22.8°C) or higher for 1,500 or more hours

For SI: °C = [(°F)-32]/1.8; 1 inch = 2.54 cm.

TABLE C301(2)
INTERNATIONAL CLIMATE ZONE DEFINITIONS

ZONE NUMBER	THERMAL CRITERIA	
	IP Units	SI Units
1	9000 <CDD50°F	5000 < CDD10°C
2	6300 < CDD50°F ≤ 9000	3500 < CDD10°C ≤ 5000
3A and 3B	4500 < CDD50°F ≤ 6300 AND HDD65°F ≤ 5400	2500 < CDD10°C ≤ 3500 AND HDD18°C ≤ 3000
4A and 4B	CDD50°F ≤ 4500 AND HDD65°F ≤ 5400	CDD10°C ≤ 2500 AND HDD18°C ≤ 3000
3C	HDD65°F ≤ 3600	HDD18°C ≤ 2000
4C	3600 < HDD65°F ≤ 5400	2000 < HDD18°C ≤ 3000
5	5400 < HDD65°F ≤ 7200	3000 < HDD18°C ≤ 4000
6	7200 < HDD65°F ≤ 9000	4000 < HDD18°C ≤ 5000
7	9000 < HDD65°F ≤ 12600	5000 < HDD18°C ≤ 7000
8	12600 < HDD65°F	7000 < HDD18°C

For SI: °C = [(°F)-32]/1.8